T0296448

Combinatorics of finite geometries

This book is an introductory text on the combinatorial theory of finite geometry. It assumes only a basic knowledge of set theory and analysis, but soon leads the student to results at the frontiers of research. It begins with an elementary combinatorial approach to finite geometries based on finite sets of points and lines, and moves into the classical work on affine and projective planes. This is followed by chapters dealing with polar spaces, partial geometries, and generalized quadrangles. The second edition contains an entirely new chapter on blocking sets in linear spaces, which highlights some of the most important applications of blocking sets - from the initial game-theoretic setting to their very recent use in cryptography. Extensive exercises at the end of each chapter ensure the usefulness of this book for senior undergraduate and beginning graduate students.

LYNN MARGARET BATTEN

Combinatorics of finite geometries

Second edition

CAMBRIDGE
UNIVERSITY PRESS

CAMBRIDGE UNIVERSITY PRESS
Cambridge, New York, Melbourne, Madrid, Cape Town, Singapore, São Paulo

Cambridge University Press
The Edinburgh Building, Cambridge CB2 2RU, UK

Published in the United States of America by Cambridge University Press, New York

www.cambridge.org
Information on this title: www.cambridge.org/9780521590143

First published 1986
Second edition 1997

A catalogue record for this publication is available from the British Library

Library of Congress Cataloguing in Publication data

Batten, Lynn Margaret.

Combinatorics of finite geometries / Lynn Margaret Batten. –
2nd ed.

p. cm.

Includes bibliographical references (p. –) and indexes.

ISBN 0-521-59014-0 (hardback). – ISBN 0-521-59993-8 (paperback)

1. Finite geometries. I. Title.
QA167.2.B38 1997
516′.13 – dc20 96-46187
CIP

ISBN-13 978-0-521-59014-3 hardback
ISBN-10 0-521-59014-0 hardback

ISBN-13 978-0-521-59993-1 paperback
ISBN-10 0-521-59993-8 paperback

Transferred to digital printing 2006

TO MY SISTER AUDREY

'I have opened all the doors in my head.'

Marilyn French, *The Women's Room*

Contents

x: blank

Preface

The principal changes to the second edition occur in sections 4.5 (which has been completely re-done) and 6.2 (which has been brought up to date). I want to thank particularly Stan Payne for his help in reviewing the appropriate literature for inclusion in 6.2.

In addition, a new chapter, chapter 8, on blocking sets, has been added. Blocking sets have numerous applications in game theory and in testing of statistical designs. We also describe some very recent applications to cryptography.

I wish to thank the University of Manitoba at this time for its support in the production of this new edition.

LMB

Preface to the first edition

This text is designed as an introduction to finite geometry for the undergraduate student. It could be used for second or third year students with some aptitude for, but not necessarily a great deal of background in, mathematics. A second year general student would have a good foundation in synthetic geometry with the completion of the first four chapters. A third year honours student or a fourth year student could be expected to complete the book in a one year course.

As far as background is concerned, only a fundamental knowledge of functions and set theory to the first year university level is essential. Some linear algebra and field theory would be useful for some parts of chapters 5, 6 and 7, but at a minimal level.

Listed in the last section of each chapter are forty to fifty exercises. Some of these are designed to consolidate the student's acquaintance with the concepts presented in the chapter. Others give additional results and introduce new concepts. The exercises form an important part of the material and I would strongly advise that students be assigned several each week. The very difficult problems have been starred.

The material presented in the book is based on the notion of 'connection number' in a near-linear space. Given a system of points and lines, for any point p not on a line ℓ, the connection number, $c(p, \ell)$, is the number of points on ℓ which are connected to p by a line. In chapter 1 (near-linear spaces) there are no restrictions on $c(p, \ell)$. In chapter 2 (linear spaces), $c(p, \ell)$ must always be the total number of points on ℓ. Chapters 3 and 4 look at special cases of the systems discussed in chapter 2. These are the classical affine and projective spaces. Chapters 5, 6 and 7 deal with structures which have been introduced more recently; in chapter 5, $c(p, \ell)$ must always be 1 or the total number of points on ℓ. Chapter 6 deals more particularly with those structures for which $c(p, \ell)$ is always 1. Finally, in chapter 7 we look at near-linear spaces for which $c(p,\ell)$ is always a fixed positive integer.

In North America, the material covered in chapters 2, 3 and 4 is generally acknowledged to be the 'classical' synthetic geometry. However, the geometries of chapters 5, 6 and 7 are already a part of the curricula of Western European schools and, in view of the important results of researchers on these geometries, it seems likely that in the near future polar spaces and partial geometries will be an intrinsic part of the North American programme.

I am indebted to many friends and colleagues for their kindness and support while this book was in progress. A special thanks to those who took time to proof-read and to make important suggestions: Francis Buekenhout, Frank de Clerck, James Hirschfeld, Stan Payne, Clare and Tom Ralston. To one of my students, Elizabeth Teitsma, a warm thank-you for providing me with diagrams of projective and affine planes, and also for proof-reading. A final thank-you to the second year Synthetic Geometry class at the University of Winnipeg, 1980–1, who put up with the disorganization that is part of putting a text together, and who supplied me with comments as to improvements of proofs, and levels of difficulty of problems.

I wish to acknowledge the University of Winnipeg for its financial support for typing of the final draft of the manuscript. And to Betty Harder who typed this final draft, my deep appreciation for the excellent and very professional finished product.

LMB

1

Near-linear spaces

I suppose you know that students of geometry and arithmetic and so forth begin by taking for granted odd and even, and the usual figures, and the three kinds of angles, and things akin to these, in every branch of study; they take them as granted and make them assumptions or postulates, and they think it unnecessary to give any further account of them to themselves or others, as being clear to everybody. Then, starting from these, go on through the rest by logical steps until they end at the object which they set out to consider.

Plato *The Republic* Book VI

1.1 Some basic concepts: consistency and dependence

The understanding throughout this book is that we work with a set P whose elements are called *points*, and a set L of certain subsets of P, whose elements are called lines.[†] We remind the reader that by definition a set has *distinct* elements.

A *space* $S=(P,L)$ is a system of points P and lines L such that certain conditions or *axioms* are satisfied. We can then consider two points of view: given a system of axioms about points and lines, can we find any spaces which satisfy it, or, given a familiar space (for example, real 3-space), what system or systems of axioms can be used to define it? We are only interested here in the former of these two questions.

As we shall be working a great deal with axiom systems, we discuss some of their properties in this section.

An axiom system is said to be *consistent* if it is possible to construct an example of a structure satisfying all the axioms. Otherwise the system is said to be *inconsistent*.

Consider the following examples where we suppose always that we are working with a system of points and lines.

Example 1.1.1

1. There are five points and six lines.
2. Each point is in one line.
3. Each line contains one point.

[†] The words *variety* and *block* are often used instead of *point* and *line* and later we shall introduce v and b for the number of points and lines respectively.

Before checking this system for consistency or inconsistency, we note the following convention. When we say there are five points, we always mean that there are *precisely* five points. Otherwise we shall add 'at least' or 'at most' or some equivalent expression.

After spending some time trying to construct an example satisfying these three axioms, it becomes apparent that we are trying to set up a one-to-one correspondence (by 2 and 3) between sets with five points and six points respectively. This of course is impossible, so the system is in fact an inconsistent one.

Example 1.1.2

1. There are seven points and seven lines.
2. Each line has three points.
3. Each point is on three lines.

The fact that one has just spent an hour trying to find an example of such a space, with no luck, does not necessarily mean that no example exists. In fact, figure 1.1.1 represents an example of just such a system. If the points are labelled 0, 1, 2, 3, 4, 5, and 6, we can choose as lines the sets $\{1, 2, 4\}$, $\{2, 3, 5\}$, $\{3, 4, 6\}$, $\{0, 4, 5\}$, $\{1, 5, 6\}$, $\{0, 2, 6\}$ and $\{0, 1, 3\}$.

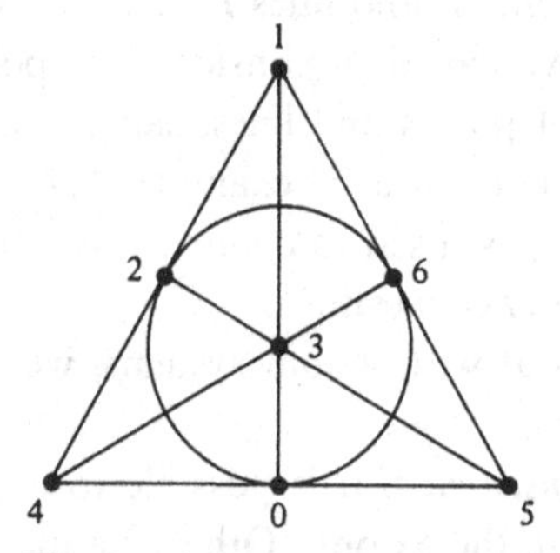

Figure 1.1.1.

Of course the labelling is arbitrary, but our choice here has an interesting property. Each line is of the form $\{1+i, 2+i, 4+i\}$ where i ranges between 0 and 6, and we always use the remainder on division by 7. That is, the set $\{1+5, 2+5, 4+5\} = \{6, 7, 9\}$ becomes the line $\{6, 0, 2\}$. This (near-linear) space is called the *Fano plane*.

So, to prove that a system is inconsistent, we must supply a formal proof of the fact. Whereas to prove consistency it suffices to provide an example.

A system of axioms is said to be *dependent* if one or more axioms can be proved using the remaining axioms. Otherwise the system is *independent*. We note that to prove dependency, it is enough to show that a particular axiom follows from the others. To prove independency, one must show that no axiom follows from the others.

Example 1.1.3

1. There are six points and four lines.
2. Each line has two points.
3. Each point is on at most four lines.

We claim that this system is dependent. In fact, it is easy to see that axiom 3 follows immediately from axiom 1: if there are only four lines, then clearly no point can be on more than four lines.

An alternative approach to proving dependence is to list all examples of systems satisfying axioms 1 and 2. If they all satisfy axiom 3, then axiom 3 follows from axioms 1 and 2.

To list all such examples, we use a systematic approach. First list all examples in which some point is on the maximum possible number of lines: four. Then list all examples for which some point is on three lines, but no point is on four lines; and so on. Diagrams representing these systems can be found in figure 1.1.2.

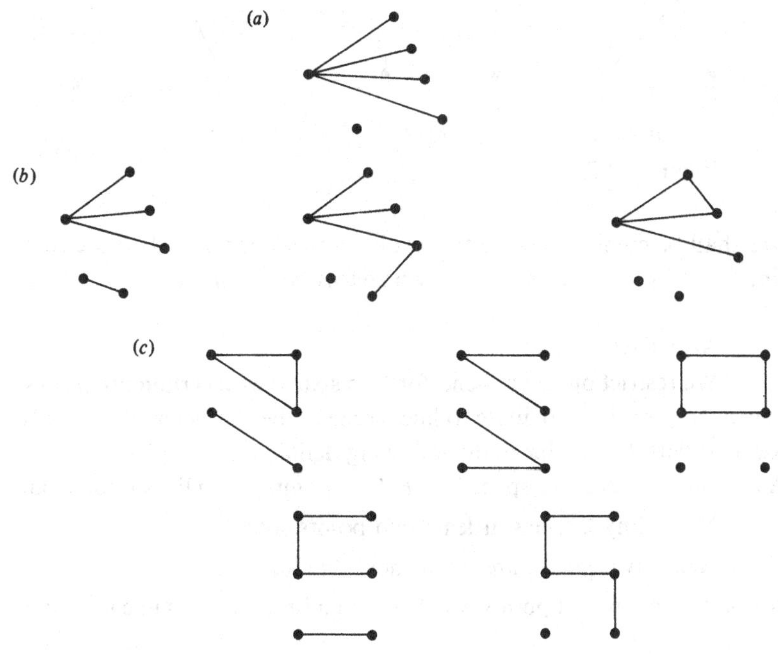

Figure 1.1.2.

Example 1.1.4

1. There are four points and four lines.
2. Every point is on two lines.

This axiom system is quickly seen to be independent, as we can provide an example of a space satisfying 1 but not 2, and an example of a space satisfying 2 but not 1.

Let $P = \{1, 2, 3, 4\}$ and $L = \{\{1, 2\}, \{2, 3\}, \{1, 3\}, \{2, 4\}\}$. This satisfies 1 but not 2.

Let $P = \{1, 2, 3\}$ and $L = \{\{1, 2\}, \{2, 3\}, \{1, 3\}\}$. This satisfies 2 but not 1.

Example 1.1.3 brings to the fore the following problem: when are two spaces the same? Intuitively we shall say that they are the same if a diagram for one can be twisted into a diagram for the other without 'breaking' any lines and without adding any extra 'joins'. Consider figure 1.1.3. The examples (*a*) and (*b*) are the same as we can move the line $\{2, 3\}$ around until the point 3 is on the other side of the line $\{1,4\}$. However, neither is the same as (*c*), because the points 3 and 4 would have to be joined together as one point.

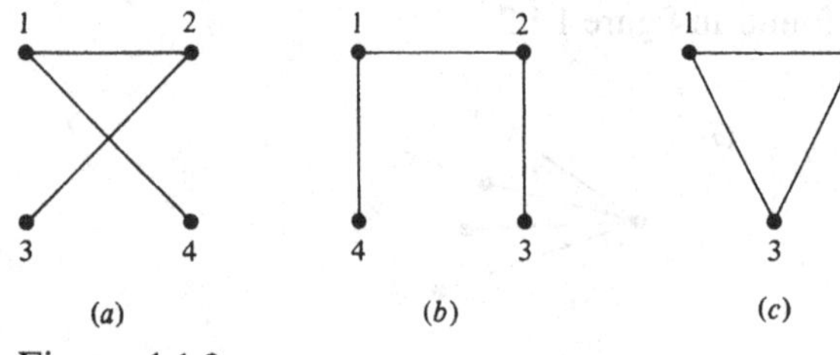

Figure 1.1.3.

We shall be coming back to this problem in section 1.7 where we shall define formally what we mean by two spaces being 'the same'.

1.2 Near-linear spaces

We restrict ourselves henceforth to spaces with certain properties. In particular, we shall attribute to lines some of the characteristics usually associated with them. Hence the following definition.

A *near-linear space*† is a space $S = (P, L)$ of points P and lines L such that

NL1 any line has at least two points, and

NL2 two points are on at most one line.

If p and q are distinct points which are on a line, then this line is unique

† Also called *partial plane*, a term introduced by M. Hall (1943).

by NL2. We denote this unique line by pq. It should be clear then, that if r and s are any distinct points on the line pq, it must be the case (by NL2) that $pq = rs$.

Example 1.2.1. Let P be the set of points of Euclidean (real) 3-space, and let L be the set of all usual lines. Then (P, L) is a near-linear space.

Example 1.2.2. Let P be as in example 1.2.1 but let L be the set of all usual planes in 3-space. This is not a near-linear space as two points are on many planes.

Example 1.2.3. Let $P = \{1, 2, 3, 4, 5, 6\}$ and $L = \{\{1, 2, 3\}, \{2, 4\}, \{3, 4, 5\}, \{1, 4\}\}$. (See figure 1.2.1.) Then (P, L) is a near-linear space.

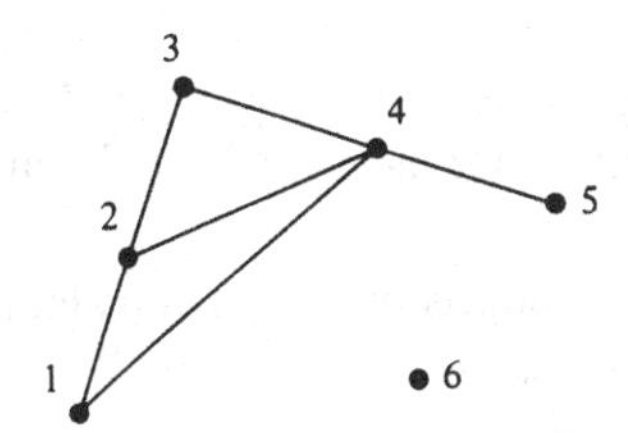

Figure 1.2.1.

Example 1.2.4. Let $P = \{1, 2, 3\}$, $L = \{\varnothing, \{1, 2\}, \{1, 2, 3\}\}$. (See figure 1.2.2 – it is rather difficult to draw the empty set.)

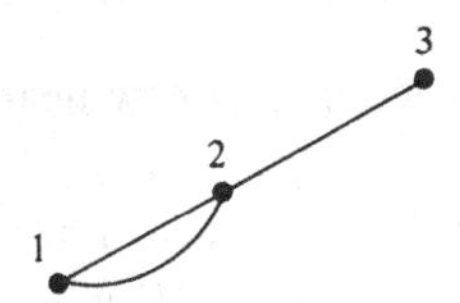

Figure 1.2.2.

This is clearly not a near-linear space. It violates both axioms in fact.

The reader can check that the spaces of figures 1.1.1, 1.1.2 and 1.1.3 are all near-linear spaces.

The definition of near-linear space does not assume that there are any points at all. In this case, there are no lines, and the near-linear space is denoted by $\varnothing$, the empty set. Furthermore, even if there are points, the definition does not imply that there are lines.

A word here about notation. In general, a point will be labelled $1, 2, \ldots$ or $a, b, \ldots$ in examples, but in proofs, $p, q, \ldots$ will be more common. To label lines, we use ℓ, and sometimes m and h. Often subscripts will be used. These are merely guidelines however, and we do not promise to stick to them entirely.

For the *order*, or number of elements of a set X, we use $|X|$.

Near-linear spaces have properties other than NL1 and NL2 in common, as we shall see by the following lemmas.

Lemma 1.2.1. *Two distinct lines of a near-linear space intersect in at most one point.*

Proof. Suppose ℓ_1 and ℓ_2 are distinct lines. If $|\ell_1 \cap \ell_2| \geq 2$, we contradict NL2. □

Lemma 1.2.2. *If ℓ_1 and ℓ_2 are such that $\ell_1 \subseteq \ell_2$, then $\ell_1 = \ell_2$.*

Proof. By NL1, ℓ_1 has at least two points, so that, by NL2, $\ell_1 = \ell_2$. □

Before moving into the next section, we introduce more notation. For the number of points in a near-linear space we use v, and for the number of lines, b.

For a line ℓ, $v(\ell)$ will be the number of points on ℓ or, equivalently, $|\ell|$.

For a point p, $b(p)$ will be the number of lines on p.

Note that these numbers may be infinite.

When talking about points and lines, we shall use the ordinary language of geometry: for example, points are *on* (rather than *in*) or *incident with* lines, lines are *on* or *incident with* points, points are *joined* by lines, etc.

1.3 New near-linear spaces from old

In this section we consider the construction of a new near-linear space from a given one.

Let $S = (P, L)$ be a near-linear space. We define a new near-linear space $R = (P', L')$ as follows. P' is an arbitrary subset of P and L' is the set of intersections $\ell \cap P'$ for any ℓ in L with at least two points in P'. It is an easy exercise to check that R is indeed a near-linear space. R is called a *restriction* of S and, in particular, the *restriction of S to P'*.

Example 1.3.1. Let P be the set of points of Euclidean 2-space, and L the usual lines. Define P' to be those points of P inside the unit circle centred at the origin. That is, a point (x, y) is in P' if $x^2 + y^2 < 1$. Any line of L meets P' in many points or misses it entirely. (The reader is asked to verify this fact.) Hence lines of L' are obtained from lines of L which meet P', restricted to the set of points in P'.

Example 1.3.2. Let $P = \{1,2,3,4,5\}$, $L = \{\{1,4\}, \{2,3\}, \{1,2,5\}\}$. If $P' = \{2,3,4\}$, then $L' = \{\{2,3\}\}$.

As the choice of P' is made arbitrarily, we get a restriction corresponding to every subset of P. Hence, if v is finite, the number of restrictions might be 2^v. (Recall that a set of v elements has 2^v subsets.) We ask, then, if all restrictions of a given space are necessarily different where, by 'different', we mean our intuitive sense of the word, described at the end of section 1.1. We answer the question in the negative by considering the following example.

Example 1.3.3. $P = \{1, 2, 3, 4, 5\}$ and each pair of points is a line. (See figure 1.3.1.) Here the 'different' restrictions are $\varnothing$, a point, a line, a triangle, a square with its diagonals, and the whole space. So there are only five new near-linear spaces obtained out of a possible $2^5 = 32$.

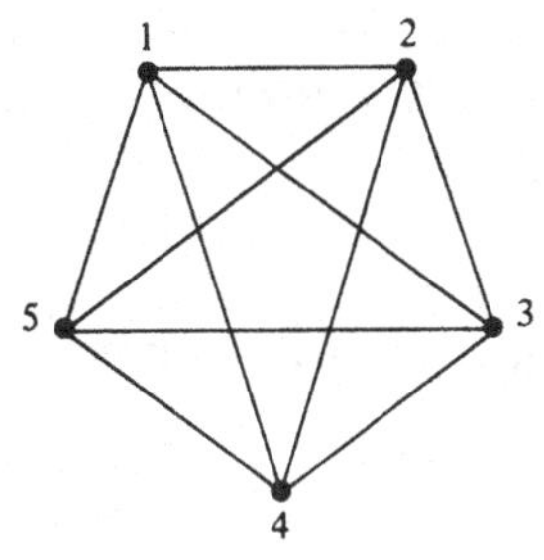

Figure 1.3.1.

Let $S = (P, L)$ be a near-linear space. We define the *dual (near-linear) space* $R = (P', L')$ of S as follows

$$P' = L$$

and any set of at least two lines of S which is the set of all lines through a fixed point of S is a line of L', and these are the only lines. In brief, $L' = \{\{p_1, \ldots, p_m\} \mid p_i \in P',\ m \geq 2$ and $p_1, \ldots, p_m$ are all the lines of S incident with a fixed point$\}$.

We illustrate this definition with the near-linear space of figure 1.2.1 where the lines may be labelled as follows: $\ell_1 = \{1, 2, 3\}$, $\ell_2 = \{3, 4, 5\}$, $\ell_3 = \{1, 4\}$, $\ell_4 = \{2, 4\}$. Then $P' = \{\ell_1, \ell_2, \ell_3, \ell_4\}$. Now for each point of S on at least two lines we obtain a line of L': $L' = \{\{\ell_1, \ell_3\}, \{\ell_1, \ell_4\}, \{\ell_1, \ell_2\}, \{\ell_2, \ell_3, \ell_4\}\}$. The dual space is illustrated in figure 1.3.2.

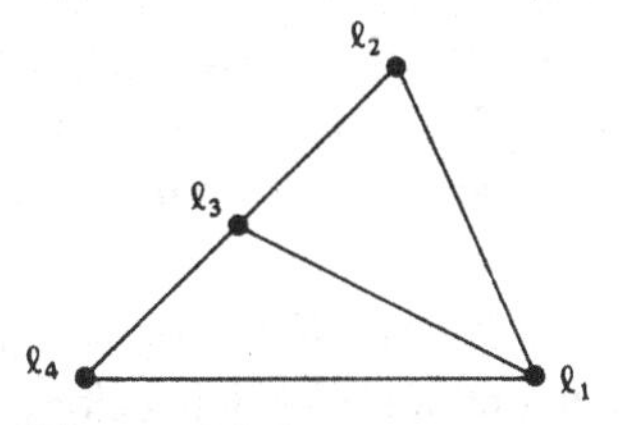

Figure 1.3.2.

Lemma 1.3.1. *The dual space of a near-linear space is a near-linear space.*

Proof. By definition, any line in the dual space has at least two points so that NL1 is satisfied.

Consider two points of the dual space and let ℓ_1 and ℓ_2 be the lines of the near-linear space $S=(P,L)$ to which these two points correspond. Each line joining ℓ_1 and ℓ_2 in the dual space corresponds to a point of intersection of ℓ_1 and ℓ_2 in S and, since there is at most one such point of intersection by lemma 1.2.1, there is at most one line on ℓ_1 and ℓ_2 in the dual space. □

A *graph* is a near-linear space in which every line has precisely two points. The near-linear spaces of figures 1.1.3 and 1.3.1 are graphs.

It is possible to obtain a graph from a near-linear space in almost the same way as we obtain the dual space. Let $S=(P,L)$ be a near-linear space; the *line graph* $R=(P',L')$ is defined by

$$P' = L,$$

$$L' = \{\{p,q\} \mid \text{ where } p \text{ and } q \text{ are distinct intersecting lines of } L\}.$$

It is clearly a graph.

The line graph of figure 1.2.1, labelling the lines as above, is shown in figure 1.3.3.

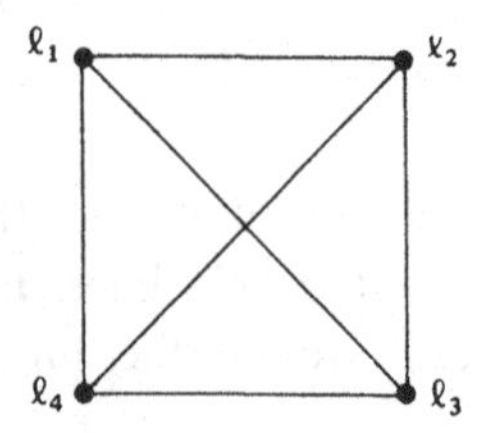

Figure 1.3.3.

The line graph of figure 1.1.2 (*a*) is also the graph of figure 1.3.3.

A *subspace* of a near-linear space (P,L) is a set X of points of P such that whenever p and q are points of X which are on a line pq of L, then the entire line pq is in X. The empty set, any point, any line and the whole space itself are always subspaces of a given space. The subspaces of figure 1.1.3 (*a*) are the ones mentioned above along with the sets $\{1,2,3\}$, $\{1,2,4\}$ $\{1,3\}$, $\{2,4\}$, $\{3,4\}$, $\{1,3,4\}$ and $\{2,3,4\}$. The subspaces of figure 1.1.1 are *only* the ones mentioned above.

A subspace becomes a near-linear space (check!) if we consider it to be the set of its points *and* of the lines of the space which belong to the subspace.

Lemma 1.3.2. *The intersection of any number of subspaces is a subspace.*

Proof. Let X be the intersection of any number of subspaces. We need only show that, if p and q are points of X and p and q are on a line pq, then $pq \subseteq X$. But any subspace containing X then contains p and q, and so by definition pq. Therefore the line pq is in all subspaces of which X is the intersection, and so pq is a subset of X. □

1.4 Dimension

Let X be any set of points of a near-linear space $S=(P,L)$. The *closure* of X is a subspace which contains X but does not properly contain any subspace on X.

It is not obvious from the definition that the closure of X is unique, but this follows from lemma 1.4.1 below. The closure of X is thus the *smallest* subspace containing X.

Notation. $\langle X \rangle$ will denote the closure of the set X.
We first of all consider some examples.
In figure 1.1.1, $\langle \{5,6\} \rangle = \{1,5,6\}$ and $\langle \{0,3,4\} \rangle = S$.
In figure 1.2.1, $\langle \{1,2,5\} \rangle = \{1,2,3,4,5\}$.

In example 1.2.1, the closure of any set of at least two collinear points is the line on them. The closure of any set of four points, no three collinear, is the whole space.

In any near-linear space S, $\langle \varnothing \rangle = \varnothing$, $\langle p \rangle = p$ and $\langle S \rangle = S$. Also, for any set of points X, $X \subseteq \langle X \rangle, \langle X \rangle = \langle \langle X \rangle \rangle$ and if $X \subseteq Y$ then $\langle X \rangle \subseteq \langle X \rangle$.

Lemma 1.4.1. *The closure of a set X is the intersection of all subspaces on X.*

Proof. By lemma 1.3.2 this intersection is a subspace. It is easy to see that it is the smallest subspace on X as any subspace on X is included when we take the intersection. □

We say that X *generates* its closure. Conversely, given a subspace R we say that X is a *generating set* for R if $\langle X \rangle = R$, so that, also, X generates R.

In section 1.1 we defined an independent axiom system as one in which no axiom followed from the others. So no superfluous information is given. In the same vein, we wish to make a definition of independence for subsets of a near-linear space. An independent set will be one which has 'just enough' points to generate its closure.

An *independent set* X is a set of points such that for each $x \in X$, $x \notin \langle X \backslash \{x\} \rangle$.

Again, let us illustrate this definition using figure 1.1.1. In the near-linear space there, the set $X=\{1,5,6\}$ is *not* independent. It contains more than enough points to generate its closure, which is itself. The points 1 and 5 would suffice:

$$6\in\langle X\backslash\{6\}\rangle.$$

We call such a set *dependent*.
The set $X=\{1,5\}$ *is* independent:

$$1\notin\langle X\backslash\{1\}\rangle=\{5\} \quad \text{and} \quad 5\notin\langle X\backslash\{5\}\rangle=\{1\}.$$

In figure 1.2.1, the sets $\{5,6\}$, $\{4,5,6\}$ and $\{2,4,5,6\}$ are independent. The set $\{1,2,4,5\}$ is dependent.

In figure 1.3.1 the set $\{1,2,3,4,5\}$ is independent.

It is not difficult to see that the empty set is independent (there are no points to check!), and that a single point is always independent. So is any pair of points.

A *basis* of a near-linear space S is an independent subset of the points of S which generates S.

A basis is not necessarily unique. The space of figure 1.1.1 has $\{1,2,0\}$ and $\{3,6,5\}$ as bases (and many more). Any basis of the space of figure 1.2.1 must contain the point 6. $\{1,2,4,6\}$ is a basis for example. There is precisely one basis for figure 1.3.1, namely $\{1,2,3,4,5\}$.

For a given near-linear space, do all bases have the same number of elements? The answer is no, as can be seen by considering the example of figure 1.4.1. The set $\{1,2,3\}$ is a basis while so is the set $\{4,5,6,7\}$.

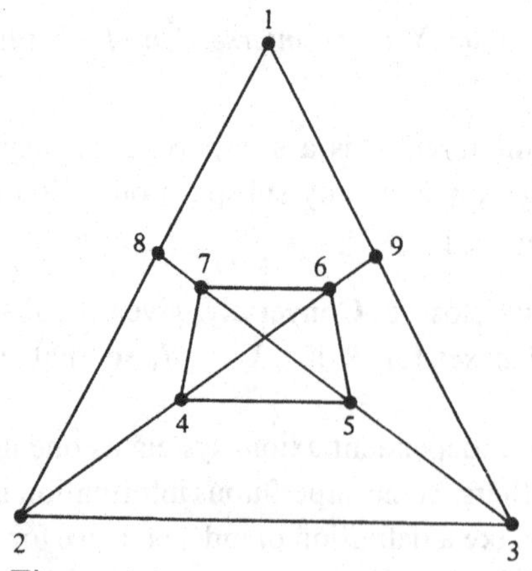

Figure 1.4.1.

We wish now to define the dimension of a finite near-linear space (i.e., one with a finite number of points) in terms of the number of elements in a basis. Moreover, thinking of examples to which we have been exposed over the years, it seems reasonable to define real '2-space' as having dimension 2, real '3-space' as having dimension 3 and so on. So we would like to define the dimension of a near-linear space to be 1 less than the number of elements in a basis. However, in view of the example of figure 1.4.1 we must make a slight adjustment.

The *dimension* of a near-linear space S is one less than $\min\{|B| \mid B$ a basis for $S\}$.[†]

In other words, we find a basis with the smallest possible number of elements and then subtract 1 from this number to get the dimension.

In chapter 2 we shall be introducing a certain class of near-linear spaces for which all bases have the same number of elements. This class has some particularly nice properties as we shall see.

Going back to the example of figure 1.4.1, we see that it has dimension 2 while so has its proper subset $\{2,4,6,7,9\}$. This seems strange indeed.

The example of figure 1.3.1 has dimension 4.

The dimension of a line, point and $\varnothing$ are respectively always 1, 0 and -1.

1.5 Incidence matrices

Any finite space (P, L) can be represented by a matrix (in fact by several). Each point corresponds to a row of the matrix and each line to a column. We put a 1 in the (i,j)th position if the point p_i is on the line ℓ_j, and otherwise we put a 0. So, for instance, a matrix for the near-linear space of figure 1.2.1 might be

$$\begin{array}{c} \\ 1\\2\\3\\4\\5\\6 \end{array}\begin{array}{c} \begin{array}{cccc} \ell_1 & \ell_2 & \ell_3 & \ell_4 \end{array} \\ \begin{pmatrix} 1 & 0 & 1 & 0 \\ 1 & 0 & 0 & 1 \\ 1 & 1 & 0 & 0 \\ 0 & 1 & 1 & 1 \\ 0 & 1 & 0 & 0 \\ 0 & 0 & 0 & 0 \end{pmatrix} \end{array}$$

where $\ell_1 = \{1,2,3\}$, $\ell_2 = \{3,4,5\}$, $\ell_3 = \{1,4\}$, $\ell_4 = \{2,4\}$.

Where we have some choice in constructing the matrix is in the labelling of the points and lines. Re-arrangement of this labelling will result in switching rows and/or columns. To restrict this choice somewhat (but not entirely), we make the following restriction. The lines will be labelled

[†] There are two schools of thought here. Some people prefer to use the maximum. See, for example, Birkhoff (1967).

$\ell_1, \ell_2, \ldots, \ell_b$ in such a way that $v(\ell_1) \geq v(\ell_2) \geq \ldots \geq v(\ell_b)$. Similarly, the points will be labelled such that $b(p_1) \geq b(p_2) \geq \ldots \geq b(p_v)$.

Fortunately, our labelling of the lines for figure 1.2.1 meets these criteria. However, the labelling of the points does not. To remedy this it suffices to exchange points 1 and 4 which results in an interchanging of rows 1 and 4 in the above matrix.

To make our notation even more concise, we define

$$v_i = v(\ell_i) \quad \text{and} \quad b_i = b(p_i).$$

Also, defining

$$r_{ij} = r(p_i, \ell_j) = \begin{cases} 0 & \text{if } p_i \notin \ell_j, \\ 1 & \text{if } p_i \in \ell_j, \end{cases}$$

we see that the (i,j)th entry of the matrix is just the number r_{ij}. The value r_{ij} is called the *incidence number* of p_i and ℓ_j, and the matrix the (an) *incidence matrix* of the space.

We can read off a great deal of information about the space just by looking at the incidence matrix. For instance, the number of 1s appearing in a row is the number of lines on the corresponding point. The number of 1s appearing in a column is the number of points on the corresponding line. Hence the point 4 is on three lines while the point 6 is on no lines in the matrix above.

If we add the 1s in each column, column by column, we get $\sum_{i=1}^{b} v_i$. If we add the 1s in each row, row by row, we get $\sum_{i=1}^{v} b_i$. But obviously we are just counting the same number of 1s in two different ways. So we have the equations

$$\sum_{i=1}^{v} r_{ij} = v_j, \quad \sum_{j=1}^{b} r_{ij} = b_i \quad \text{and}$$

$$\sum_{j=1}^{b} v_j = \sum_{j=1}^{b} \sum_{i=1}^{v} r_{ij} = \sum_{i=1}^{v} \sum_{j=1}^{b} r_{ij} = \sum_{i=1}^{v} b_i.$$

Given any matrix of 0s and 1s, when does it represent a near-linear space? We must be able to check NL1 and NL2. Assuming that the rows of the matrix correspond to the points, and the columns to the lines, NL1 says that any column has at least two 1s. NL2 says that there is at most one k such that r_{ik} and r_{jk} are both 1 for distinct i and j.

Let $S = (P, L)$ be a near-linear space and consider its dual near-linear space $R = (P', L')$. How do the incidence matrices of these two spaces compare with each other? The lines of S become the points of P', but the lines of L' correspond only to those points of P which have at least two points on them. So we eliminate in the example of figure 1.2.1 the points

5 and 6. The matrix of the dual space can be seen to be

$$\begin{array}{c} \\ \ell_1 \\ \ell_2 \\ \ell_3 \\ \ell_4 \end{array} \begin{array}{c} \begin{array}{cccc} 1 & 2 & 3 & 4 \end{array} \\ \begin{pmatrix} 1 & 1 & 1 & 0 \\ 0 & 0 & 1 & 1 \\ 1 & 0 & 0 & 1 \\ 0 & 1 & 0 & 1 \end{pmatrix} \end{array}.$$

Compare this with figure 1.3.2, the diagram of the dual near-linear space. The matrix here is the *transpose* of the matrix above except for the rows 5 and 6. That is, the r_{ji}th value in this matrix is the r_{ij}th value in the above matrix. Rows and columns of the first matrix become respectively columns and rows of the second.

We summarize some of the above in the next lemma.

Lemma 1.5.1. *If each point of a near-linear space is on at least two lines, then the number of points and lines is respectively the number of lines and points of the dual space. Moreover, since* $(M^t)^t = M$ *(the transpose of the transpose of the matrix is the matrix again), the dual of the dual of S is the original space S again (up to isomorphism – see section 1.7).*

We do point out that the lines and points obtained in the dual space are not necessarily ordered in the way we explained earlier.

Any subspace of a near-linear space is itself a subspace, as we mentioned above. Moreover, if we order the points and lines in the whole space as above, then the ordering induced in the subspace is also of the type we want. So an incidence matrix of the subspace with the proper ordering is obtained merely by taking those rows corresponding to the points of the subspace, intersected with those columns having at least two points in the subspace, and writing these down in the same order. From the subspace $\{3,4,5\}$ of figure 1.2.1 we obtain the matrix

$$\begin{array}{c} \\ 3 \\ 4 \\ 5 \end{array} \begin{array}{c} \ell_2 \\ \begin{pmatrix} 1 \\ 1 \\ 1 \end{pmatrix} \end{array}.$$

From the subspace $\{1,2,3,4,5\}$ we obtain

$$\begin{array}{c} \\ 1 \\ 2 \\ 3 \\ 4 \\ 5 \end{array} \begin{array}{c} \begin{array}{cccc} \ell_1 & \ell_2 & \ell_3 & \ell_4 \end{array} \\ \begin{pmatrix} 1 & 0 & 1 & 0 \\ 1 & 0 & 0 & 1 \\ 1 & 1 & 0 & 0 \\ 0 & 1 & 1 & 1 \\ 0 & 1 & 0 & 0 \end{pmatrix} \end{array}.$$

Note that if we choose a set of points which includes no lines (for example $\{1\}$ or $\{2,6\}$ in the above), we do not get a matrix.

A near-linear space is said to be *line regular* if $b \geq 1$ and each line has the same number of points. This number is called the *line regularity*. The space is *point regular* if $v \geq 1$ and each point is on the same number of lines and this number is called the *point regularity*.

Lemma 1.5.2. *Let S have line regularity s and point regularity t. Then* $vt = bs$.

Proof. This follows immediately from consideration of the incidence matrix. □

So for a near-linear space with v points and b lines, which is known to be both point regular and line regular, the point regularity determines the line regularity and vice versa.

A line with k points will be called a *k-line*. A point with k lines is a *k-point*.

1.6 The connection number

We assume in this section that we are working with a finite near-linear space.

Let p_i be a point not on the line ℓ_j. We define the *connection number* $c(p_i, \ell_j) = c_{ij}$ to be the number of points of ℓ_j joined to p_i by a line. Because of NL2, this is the same as the number of lines on p_i meeting points of ℓ_j. If $p_i \in \ell_j$, can we define the connection number? We have a choice here. It could be the number of points of ℓ_j different from p_i, or the number of lines connecting p_i and points of ℓ_j, namely 1. We use the latter definition. Hence, $c_{ij} = 1$ if $r_{ij} = 1$.

Lemma 1.6.1. *For any point* p_i *and line* ℓ_j, $c_{ij} \leq v_j$.

Proof. This follows easily from NL2 if $r_{ij} = 0$. If $r_{ij} = 1$ it follows from NL1. □

Lemma 1.6.2. *If* $r_{ij} = 0$ *then the number of lines on* p_i *missing* ℓ_j *is* $b_i - c_{ij}$.

Proof. b_i is the total number of lines on p_i and c_{ij} is the number of these which meet ℓ_j. Hence the result is immediate. □

We define a *linear space* to be a near-linear space in which any two points are on a line. The examples of figures 1.1.1, 1.1.3 (c), 1.3.1, 1.3.2 and 1.3.3 are all linear spaces.

It is immediate that in a linear space, $r_{ij} = 0$ implies $c_{ij} = v_j$.

Lemma 1.6.3. *If S is a near-linear space with* $b \geq 1$ *and if* $c_{ij} = v_j$ *for every* p_i *and* ℓ_j *such that* $r_{ij} = 0$, *then S is a linear space.*

Proof. Since $b \geq 1$, there is a line ℓ_k, say. We must show that any two points are on a line. Let p_i and p_j be two points. If $r_{ik} = r_{jk} = 1$, we are done. If $r_{ik} = 0$ and $r_{jk} = 1$ then, by assumption, $c_{ik} = v_k$ so that p_i is joined to every point of ℓ_k by a line. In particular, p_i is joined to p_j.

Finally, if $r_{ik} = r_{jk} = 0$, we have a line p_iq where $q \in \ell_k$, using the hypothesis once again. If $p_j \in p_iq$ we are finished and, otherwise, apply the hypothesis one last time to get a line p_ip_j. □

The assumption that S have at least one line is essential since a set of at least two points with no lines satisfies the hypothesis (vacuously, i.e., there are no lines to check!), while S is certainly not a linear space.

The next result is an important enough result for us to call it a theorem.

Theorem 1.6.4. *Let $S = (P, L)$ be a finite near-linear space with v points and b lines. Then S is a linear space if and only if $\sum_{j=1}^{b} v_j(v_j - 1) \geq v(v-1)$. (A sum with no entries is assumed to be zero.)*

Proof. Suppose that S is a linear space. We count the number of pairs of points in two different ways. First of all, there are $\binom{v}{2}$ pairs of points (counting $\{p_i, p_j\}$ to be the same pair as $\{p_j, p_i\}$), or $v(v-1)/2$. Also, as any pair of points determines a unique line, the total number of pairs of points is the total number of pairs of points on each line, summed over all lines, that is, $\sum_{j=1}^{b} v_j(v_j - 1)/2$.

So

$$\sum_{j=1}^{b} v_j(v_j - 1) = v(v-1).$$

In fact we have equality here!

Suppose, conversely, that $\sum_{j=1}^{b} v_j(v_j - 1) \geq v(v-1)$. We prove that S is linear by induction[†] on v.

As $\varnothing$ and a single point are trivially linear spaces, we may as well assume that $v \geq 2$. If $v = 2$, there are two possibilities: $b = 0$ or 1. The inequality above only holds, however, when $b = 1$ and so S is a linear space. If $v = 3$, there are precisely four possibilities for b: $b = 0$, 1, 2 or 3. Of these, only the cases $b = 1$, $v_1 = v = 3$, and $b = 3$, $v_1 = v_2 = v_3 = 2$ satisfy the above inequality. In both of these cases, S is a linear space.

Suppose then that if the inequality holds for a near-linear space S' with fewer than v points then S' is a linear space. We must consider our near-linear space S with v points. We may assume $\sum_{j=1}^{b} v_j(v_j - 1) \geq v(v-1)$ in S, where $v \geq 4$[‡]. Let p be a point of S and consider the near-linear space $R = (P', L')$ which is the restriction of S to $P \backslash \{p\}$. So the points of R are the

[†] For some information on proofs by induction, see Curtis (1967).

[‡] In fact, 3 would suffice.

points of $P\backslash\{p\}$ and the lines of R are those lines of L not on p along with those lines on p of order at least 3, but without the point p.

As $|P'| = v-1 < v$, we attempt to prove that R is a linear space by showing that the appropriate inequality above holds. Its right hand side becomes $(v-1)(v-2)$.

For any line of L' which was ℓ_j in L, we use ℓ'_j, and the corresponding number of points is v'_j. Working in R, we have

$$\begin{aligned}\sum_{\ell'_j} v'_j(v'_j-1) &= \sum_{\substack{\ell_j \text{ not}\\ \text{on } p}} v_j(v_j-1) + \sum_{\substack{\ell_j \text{ on } p\\ |\ell_j|\geqq 3}} v'_j(v'_j-1)\\ &= \sum_{\substack{\ell_j \text{ not}\\ \text{on } p}} v_j(v_j-1) + \sum_{\substack{\ell_j \text{ on } p\\ |\ell_j|\geqq 3}} (v_j-1)(v_j-2)\\ &= \sum_{\substack{\ell_j \text{ not}\\ \text{on } p}} v_j(v_j-1) + \sum_{\substack{\ell_j \text{ on } p\\ |\ell_j|\geqq 3}} v_j(v_j-1) - 2\left(\sum_{\substack{\ell_j \text{ on } p\\ |\ell_j|\geqq 3}} (v_j-1)\right).\end{aligned}$$

In S,

$$\sum_{\ell_j} v_j(v_j-1) = \sum_{\substack{\ell_j \text{ not}\\ \text{on } p}} v_j(v_j-1) + \sum_{\substack{\ell_j \text{ on } p\\ |\ell_j|\geqq 3}} v_j(v_j-1) + \sum_{\substack{\ell_j \text{ on } p\\ |\ell_j|=2}} v_j(v_j-1).$$

So

$$\sum_{\substack{\ell_j \text{not}\\ \text{on} p}} v_j(v_j-1) = \sum_{\ell_j} v_j(v_j-1) - \sum_{\substack{\ell_j \text{ on } p\\ |\ell_j|\geqq 3}} v_j(v_j-1) - \sum_{\substack{\ell_j \text{ on } p\\ |\ell_j|=2}} v_j(v_j-1).$$

Substituting in the above, we get

$$\begin{aligned}\sum_{\ell'_j} v'_j(v'_j-1) &= \sum_{\ell_j} v_j(v_j-1) - 2\sum_{\substack{\ell_j \text{ on } p\\ |\ell_j|\geqq 3}} (v_j-1) - \sum_{\substack{\ell_j \text{ on } p\\ |\ell_j|=2}} v_j(v_j-1)\\ &= \sum_{\ell_j} v_j(v_j-1) - 2\left(\sum_{\ell_j \text{ on } p} (v_j-1)\right).\end{aligned}$$

By hypothesis, $\sum_{\ell_j} v_j(v_j-1) \geq v(v-1)$. Moreover, by counting the points on the lines on p, it becomes evident that $\sum_{\ell_j \text{on} p}(v_j-1) \leq v-1$ and so $-2(\sum_{\ell_j \text{ on } p}(v_j-1)) \geq -2(v-1)$. Therefore

$$\sum_{\ell'_j} v'_j(v'_j-1) \geq v(v-1) - 2(v-1) = (v-1)(v-2) \text{ as desired.}$$

Now, by our induction hypothesis, R is a linear space.

It remains only to show that p is joined to each point q of R. Fix q then, in R, and now choose arbitrarily a third point r in S. But $P\backslash\{r\}$ is a restriction of P with $v-1$ points and the argument used above shows that it is a linear space. Hence there is a line on p and q.

Corollary. *S is a linear space if and only if*

$$\sum_{j=1}^{b} v_j(v_j - 1) = v(v-1).$$

Proof. Recall that in the first part of the previous proof, we proved that, if S is linear, then in fact we do have equality. On the other hand, the equality certainly implies the inequality which, by theorem 6.4, implies that S is linear. □

The seemingly stronger corollary follows immediately from the theorem, and one wonders why we didn't prove it directly. In fact the induction step has an inequality at one point only which is difficult to get rid of. Can you find it?

1.7 Linear functions

Let $S = (P,L)$ and $S' = (P',L')$ be near-linear spaces. Let f be a function with domain P mapping into P'. f is a *linear function* if $f(\ell) \in L'$ for all $\ell \in L$.

A linear function is *1–1* (*one-to-one*) and/or *onto* if as a function from P to P' it is 1–1 and/or onto.

We note that if a line $\ell \in L$ is finite and if $f(\ell) \in L'$ then $v(\ell) \geq v(f(\ell))$. Hence lines may map to 'shorter' lines but not to 'longer' lines.

Let S be the near-linear space of figure 1.1.1 and S' the near-linear space of figure 1.7.1 and define f by

$f: 0 \to d$	So $f: \{1,2,4\} \to \{b,d\}$
$1 \to b$	$\{0,4,5\} \to \{d,g\}$
$2 \to b$	$\{1,5,6\} \to \{b,g\}$
$3 \to b$	$\{0,1,3\} \to \{b,d\}$
$4 \to d$	$\{2,3,5\} \to \{b,g\}$
$5 \to g$	$\{3,4,6\} \to \{b,d\}$
$6 \to b.$	$\{0,2,6\} \to \{b,d\}.$

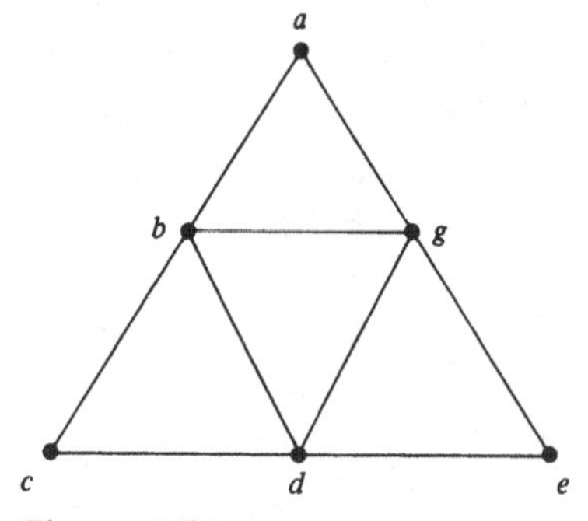

Figure 1.7.1.

Thus f is a linear function. It is obviously neither 1–1 nor onto.

We say that two near-linear spaces are *isomorphic* if there is a 1–1 linear function from one onto the other such that f^{-1} is also linear. f is called an *isomorphism*.

Recall that in section 1.1 we promised to give a more precise definition of the concept of 'sameness'. We do so here. Two near-linear spaces are *the same* if they are isomorphic and *different* otherwise. In figure 1.1.3 then, (*a*) and (*b*) are isomorphic, or the same, because we can define an isomorphism between them. For example, let $f(1)=1$, $f(2)=2$, $f(3)=3$ and $f(4)=4$. It is quickly checked that f is linear, and it is easily seen to be 1–1 and onto. If two finite sets are isomorphic then they have the same number of elements, so it follows immediately that (*c*) is not isomorphic to (*a*). It is easy to see the following.

Lemma 1.7.1. *If the linear function f from $S=(P,L)$ to $S'=(P',L')$ is 1–1, then $v(\ell)=v(f(\ell))$ for all ℓ.*

Lemma 1.7.2. *If f is an isomorphism from $S=(P,L)$ to $S'=(P',L')$ then $v(\ell)=v(f(\ell))$ and $b(p)=b(f(p))$ for all ℓ and p.*

Proof. The first part is lemma 1.7.1. For the second part, we note first that if $p\in\ell$ then $f(p)\in f(\ell)$ so that $b(f(p))\geq b(p)$. Since f^{-1} is an isomorphism from S' to S, it follows similarly that $b(p)\geq b(f(p))$. □

An *automorphism* of $S=(P,L)$ is an isomorphism from S onto itself. In geometry this is more often called a *collineation* and this is the word we shall use.

The near-linear space of figure 1.1.1 has 168 collineations.[†] Some of these are given by $f_1:(1,2,4,0,5,6,3)\to(1,2,4,0,5,6,3)$ (the ith point mapping to the ith point); $f_2:(1,2,4,0,5,6,3)\to(1,6,5,0,4,2,3)$; f_3: $(1,2,4,0,5,6,3)\to(4,0,5,6,1,2,3)$.

To find all collineations of the near-linear space of figure 1.7.1, we proceed as follows. In view of lemma 1.7.2, any collineation must map the set $\{a,c,e\}$ to itself, and the set $\{b,d,g\}$ to itself. If $f:a\to a$, there are two possibilities: $c\to e$ and $e\to c$ or $c\to c$ and $e\to e$. From the latter, since f is linear, b, d and g are all fixed and we get f_1 of figure 1.7.2.

$f_1:a\to a$	$f_2:a\to a$	$f_3:a\to c$	$f_4:a\to c$	$f_5:a\to e$	$f_6:a\to e$
$b\to b$	$b\to g$	$b\to b$	$b\to d$	$b\to g$	$b\to d$
$c\to c$	$c\to e$	$c\to a$	$c\to e$	$c\to a$	$c\to c$
$d\to d$	$d\to d$	$d\to g$	$d\to g$	$d\to b$	$d\to b$
$e\to e$	$e\to c$	$e\to e$	$e\to a$	$e\to c$	$e\to a$
$g\to g$	$g\to b$	$g\to d$	$g\to b$	$g\to d$	$g\to g$

Figure 1.7.2.

[†] See, for example, Buekenhout and Doignon (1978), pages 15 and 16.

If we let $c \to e$ and $e \to c$ we see that f linear forces $f = f_2$ of figure 1.7.2. So suppose f moves a. There are two choices. Let $f: a \to c$. If $c \to a$ then $e \to e$ and f linear implies that $f = f_3$ of figure 1.7.2. If $c \to e$, then $e \to a$ and f linear forces $f = f_4$. Finally, let $f: a \to e$. In case $c \to a$, then $e \to c$ and f must be f_5. If $f: c \to c$ then $e \to a$ and f is f_6. These are all possible collineations of S.

Notice that the identity function $f(x) = x$ for all $x \in P$ is always a collineation of $S = (P, L)$.

Lemma 1.7.3. *If f and g are collineations of $S = (P, L)$ then so are f^{-1} and $f \circ g$.*

Proof. We know that f^{-1} and $f \circ g$ are 1–1 and onto maps from P to P[†] and that f^{-1} is linear. So clearly it suffices to show that $f \circ g$ is linear. Let ℓ be a line of L. Then $g(\ell)$ is a line of L and $f(g(\ell))$ is a line of L. Thus $f \circ g$ is linear. □

A *group* is a non-empty set G with a binary operation $*$ such that

1. $a, b \in G$ implies $a * b \in G$ for all $a, b \in G$,
2. $a, b, c \in G$ implies $a * (b * c) = (a * b) * c$ for all a, b, c in G,
3. there is an $e \in G$ such that $e * a = a * e = a$ for all $a \in G$,
4. for each $a \in G$ there is a $b \in G$ such that $a * b = b * a = e$.

The element e of 3 is called an *identity* and the element b of 4 is called an *inverse* for a. Property 1 is called the *closure* property and property 2 the *associative* property. If we choose the operation $*$ to be composition of functions, and G to be the set of all collineations of a near-linear space, then the above discussion has shown that 1 and 4 hold. The identity function plays the role of e, and property 2 is clearly seen to be true.

Lemma 1.7.4. *The set of all collineations of a near-linear space is a group under composition of functions.*

If S and S' are finite isomorphic near-linear spaces, then they have the same number of points. On the other hand, it is easy to construct finite near-linear spaces with the same number of points which are not isomorphic. We ask if it is possible to give conditions on the parameters (v, b, v_i, b_i, etc.) of two near-linear spaces, to ensure that they are isomorphic. In fact, this is a very difficult question. We refer the reader to exercises 7, 9, 10 and 32 of section 1.8.

1.8 Exercises

The exercises are very generally arranged according to section.

1. Is the following axiom system of a near-linear space consistent/dependent?
   (*a*) There are six points.

[†] See, for example, Swokowski (1979).

(*b*) There are three 3-point lines.
(*c*) There are at least three 2-point lines.

2. Check for dependence and consistency: a *design* is a space (P, L) of points and lines such that
   (*a*) each line contains at least two points;
   (*b*) two points are on at least one line.
3. Find an axiom system which is both inconsistent and dependent.
4. Find an example of a near-linear space with an infinite number of points and a finite number of lines.
5. In a near-linear space with v points in which each line has three points, what is the maximum number of lines on any point? Explain.
6. If a near-linear space has v points, what is the maximum number of lines it can have? Explain.
7. Find all line regular near-linear spaces with ten points and line regularity 5.
8. (*a*) If a near-linear space is line regular, must it be point regular?
   (*b*) If it is point regular, must it be line regular?
9. Find all point regular graphs with five points.
10. (*a*) Find all near-linear spaces with four points.
    (*b*) Find all linear spaces with four points.
    (*c*) Find all graphs with four points.
11. A *complete graph* is a graph which is a linear space. Prove that, in a finite complete graph, $2b + v = v^2$.
12. Find all possible non-isomorphic near-linear spaces which arise as restrictions to four points of the space of figure 1.4.1.
13. Determine the dual space of the near-linear space of figure 1.7.1.
14. Determine which of the near-linear spaces of figure 1.1.2 have the same line graphs.
15. Find a near-linear space which is its own dual space.
16. Find the dual of the dual of figure 1.8.1.

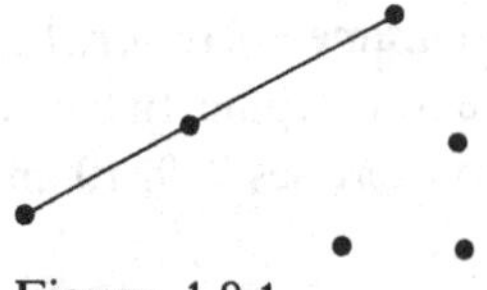

Figure 1.8.1.

17. Find a near-linear space which has the graph of figure 1.8.2 as its line graph.

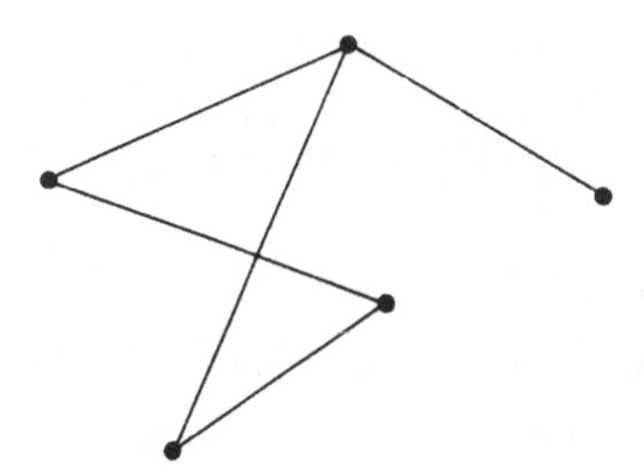

Figure 1.8.2.

18. Show that closure in a near-linear space satisfies the following properties: $X \subseteq \langle X \rangle, \langle X \rangle = \langle \langle X \rangle \rangle$, and $X \subseteq Y$ implies $\langle X \rangle \subseteq \langle Y \rangle$.

19. Find a finite near-linear space with dimension 3.

20. Find an infinite near-linear space with dimension 4.

21. Find a basis of the near-linear space of figure 1.8.3. What is its dimension? Find all subspaces.

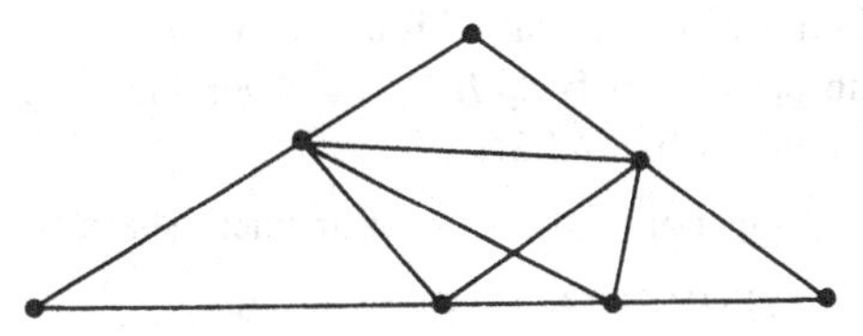

Figure 1.8.3.

22. Find a near-linear space of dimension 3 which contains a proper subspace of dimension 3. (Hint: examine figure 1.7.1.)

23. Construct the adjacency matrix of the near-linear space of figure 1.1.2 (*c*).

24. Draw the space which has the incidence matrix below. Is it a near-linear space?

$$\begin{pmatrix} 0 & 1 & 1 & 0 \\ 1 & 0 & 1 & 1 \\ 1 & 1 & 1 & 0 \\ 0 & 0 & 0 & 0 \end{pmatrix}$$

25. Is it possible to have a linear function mapping the space of figure 1.7.1 to that of figure 1.1.1?

26. Find an example of two *different* near-linear spaces with a 1–1 linear function of one onto the other.

27. Find all collineations of the near-linear space of figure 1.8.4.

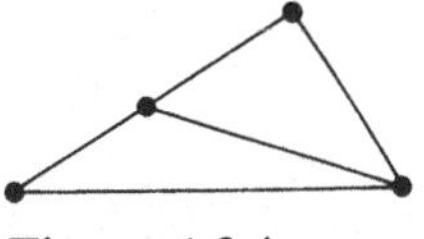

Figure 1.8.4.

28. Let $S=(P,L)$ where $P=\{1,2,3,4,5\}$ and $L=\{\{1,2\},\{1,4\},\{4,5\},\{2,5\},\{2,3,4\},\{1,3,5\}\}$, and let $S'=(P',L')$ where $P'=\{a,b,c,d\}$, $L'=\{\{a,b\},\{a,c\},\{a,d\},\{b,c,d\}\}$. Define f: $1\to a$, $2\to b$, $3\to c$, $4\to d$, $5\to a$. Is f a linear function?

29. Are the spaces S and S' of exercise 28 isomorphic?

30. Give an example of a near-linear space with both finite and infinite lines.

31. Find two infinite near-linear spaces which are isomorphic.

***32.** Show that if S and S' are near-linear spaces with the same number of points, and of lines, and if there are orderings of the v_is and b_js and v_i's and b_j's such that $v_i=v_i'$ and $b_j=b_j'$ for all i and j, then this does not necessarily imply that S and S' are isomorphic.

33. If V is a subspace of the near-linear space S and f is a linear function from S to S', show that $f(V)$ is not necessarily a subspace of S'.

34. Find an example satisfying the conditions of lemma 1.7.1 but for which $b(p)\neq b(f(p))$.

35. A *hyperplane* H of a near-linear space S is a proper subspace of S such that there is no subspace V satisfying $H\subsetneq V\subsetneq S$. Find the hyperplanes of the near-linear space of figure 1.8.3.

36. Show that $v\geq b\geq 1$ does not imply that a near-linear space is linear.

37. Check the following systems to see which are groups:
(a) the integers under addition;
(b) the integers modulo 7 under addition.

38. Construct a near-linear space with ten points and with the near-linear space of figure 1.7.1 as a subspace.

39. Find a near-linear space for which $r_{ij}=0$ implies that c_{ij} is a fixed constant, for any i and j.

40. Find the composition $f_2\circ f_3$ of f_2 and f_3 of figure 1.1.1 (see section 1.7).

41. If ℓ_1 and ℓ_2 are lines of a near-linear space which do not meet, show that the number of lines meeting both ℓ_1 and ℓ_2 is less than or equal to $v_1\cdot v_2$.

***42.** Find a near-linear space different from that of figure 1.4.1 which has bases with different numbers of elements.

2

Linear spaces

For it is but a step, saith the wise man, from the Sublime unto the Ridiculous.
And the simple dwelleth midway between, and shareth the qualities of either.

Lewis Carroll *The New Belfry of Christ Church Oxford*

2.1 Examples

We met the notion of linear space in section 1.6 but redefine it here.

A *linear space* is a space (P, L) of points and lines such that

L1 any line has at least two points, and

L2 two points are on precisely one line.

As for near-linear spaces, we use the notation pq for the line on the distinct points p and q. Clearly $r_{ij} = 0$ implies $c_{ij} = v_j$ and $b_i \geq v_j$. A linear space is called *trivial* if $b \leq 1$.

As a linear space is also a near-linear space, all the results of chapter 1 are valid here. Moreover, to convert any of our near-linear spaces of chapter 1 into linear spaces, if suffices to add as lines all pairs of points not already on a line. In particular, the space of figure 1.4.1 becomes that of figure 2.1.1.

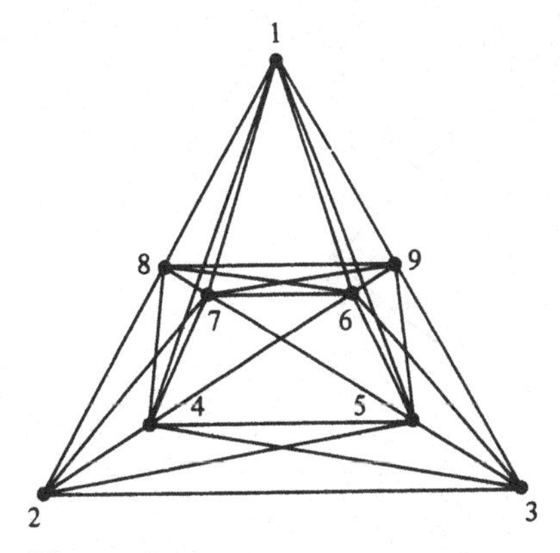

Figure 2.1.1.

It is still the case that $\{1, 2, 3\}$ and $\{4, 5, 6, 7\}$ are both bases for this space.

Example 2.1.1. Let $\mathbb{R}^2$ be the Euclidean plane. A *point* of $\mathbb{R}^2$ is an ordered pair (x, y) of real numbers. A *line* is the set of points (x, y) satisfying an equation $y = ax + b$ or $x = c$ where a, b and c are fixed real numbers. This is a linear space.

Now let $P' = \{(x, y) | x^2 + y^2 < 1\}$. That is, P' is the set of points inside a unit circle. Let L' be the set of restrictions of lines of $\mathbb{R}^2$ to P'. Since tangent lines to the circle $x^2 + y^2 = 1$ of $\mathbb{R}^2$ do not meet P' at all, any line of L' has at least two points. Clearly any two points of P' are on a unique line. Hence (P', L') is a linear space.

Example 2.1.2 *A Steiner*[†] *system* $S(t, k, v)$ is a set S of v elements called *points* in which certain subsets called *blocks* are distinguished such that

(i) any set of t distinct points ($t \geq 2$) is contained in one and only one block;
(ii) each block has exactly k points.

We are only interested in the case where 'blocks' are considered to be 'lines' and $t = 2$,[‡] that is, when the Steiner system is a linear space. If there are no blocks, (i) and (ii) are trivially satisfied. If there is one block, $k = v$.

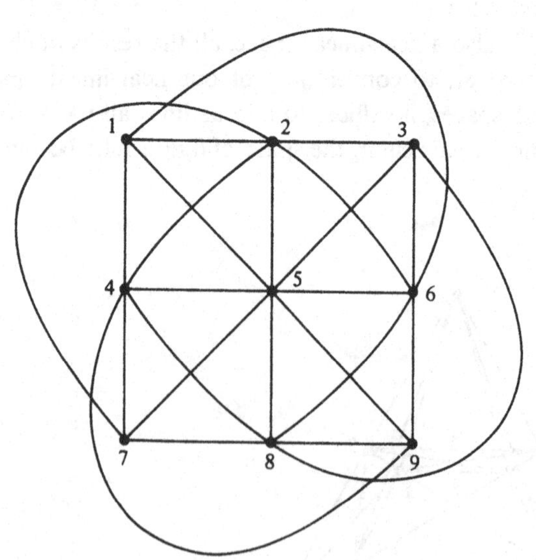

Figure 2.1.2.

[†] After Steiner (1853).
[‡] The construction of $S(t, k, v)$'s is in general difficult. For more information about this see M. Hall (1967 and 1986).

These are called *trivial* Steiner systems.

An example of an $S(2,2,3)$ is a triangle (three 2-points and three 2-lines). In fact it is easy to see that this is the unique $S(2,2,3)$. Generalizing, the only $S(2,2,v)$ is the complete graph on v points. The Fano plane (see figure 1.1.1) is an $S(2,3,7)$. The linear space of figure 2.1.2 is an $S(2,3,9)$. Apart from the obviously 'straight' lines, the four sets $\{1,6,8\}$, $\{3,4,8\}$, $\{2,6,7\}$ and $\{2,4,9\}$ are lines. Notice that this linear space has point regularity 4 and line regularity 3. Moreover, it satisfies the Euclidean 'parallel postulate': given any line ℓ and point $p \notin \ell$, there is precisely one line on p missing ℓ. (This postulate does not hold in the space (P', L') of example 2.1.1.)

Example 2.1.3. Consider the Euclidean plane (P, L) again, as in example 2.1.1. For any line ℓ let $[\ell]$ be the set of all lines parallel to ℓ including ℓ itself. We have the following properties:

(*a*) if $\ell' \in [\ell]$, then $[\ell'] = [\ell]$;
(*b*) if $\ell' \notin [\ell]$, then $[\ell'] \cap [\ell] = \emptyset$.

We construct a new system (P', L'):

$$P' = P \cup \{[\ell] \mid \ell \in L\};$$
$$L' = \{\{\ell \cup [\ell] \mid \ell \in L\},\ \{[\ell] \mid \ell \in L\}\}.$$

In other words, the points of P' are those of P along with the parallel classes $[\ell]$. These classes will be called *points at infinity*. Each line of L gets a point at infinity added to it to make it a line of L'. L' gets in addition one new line, consisting of all the new points. We call this the *line at infinity*.

Clearly, any line of L' has at least two points (in fact an infinite number). Any two points of P are on a unique line in L'. Two points at infinity are on a unique line: the line at infinity. For a point p of P and a point at infinity $[\ell]$, there is a unique line of $[\ell]$ on p. Hence, (P', L') is a linear space. It is called the *extended real plane* or the *real projective plane*.

2.2 The de Bruijn–Erdös theorem

We want to find some relationship between b and v, the number of lines and the number of points in a linear space. Checking the examples we have so far, we see that if v is finite then $b \geq v$ unless there is no or one line. In theorem 2.2.2 below, we prove that this is always true. We need the following lemma first.

Lemma 2.2.1. *In a near-linear space,* $\sum_{i=1}^{v} b_i(b_i - 1) \leq b(b-1)$.

Proof. There are $b_i(b_i - 1)$ ordered pairs of lines on the point p_i. So the left hand side of the inequality counts the number of ordered pairs of intersecting lines. Clearly there are altogether $b(b-1)$ ordered pairs of lines. □

Corollary. *In a near-linear space, any two lines meet if and only if* $\sum_{i=1}^{v} b_i(b_i-1) = b(b-1)$.

Theorem 2.2.2 (*de Bruijn–Erdös*).† *Let S be any finite linear space with* $b>1$. *Then*

(i) $b \geq v$,

(ii) *if* $b=v$, *any two lines have a point in common.*

In case (ii), either one line has $v-1$ *points and all others have two points, or every line has* $k+1$ *points and every point is on* $k+1$ *lines,* $k \geq 2$.

Proof. Let $m=\min\{b_i \mid 1 \leq i \leq v\}$. Note that $m>1$ as otherwise all points would be collinear. Let p_v be a point with m lines on it and label these lines $\ell_1, \ldots, \ell_m$ (this may not be the labelling of section 1.5). Let $p_i \in \ell_i$, $1 \leq i \leq m$, $p_i \neq p_v$. Then $p_i \notin \ell_j$, $i \neq j$. Hence $b_i \geq v_{i+1}$, $1 \leq i \leq m-1$ and $b_m \geq v_1$.

As b_v is minimal we obtain $b(p) \geq b_v \geq v(\ell)$ for all p, and for all ℓ not on p_v.

From section 1.5, $\sum_{j=1}^{b} v_j = \sum_{i=1}^{v} b_i$, so that

$$\sum_{j=1}^{m} v_j + \sum_{j=m+1}^{b} v_j = \sum_{i=1}^{m} b_i + \sum_{i=m+1}^{v} b_i$$

and, from above,

$$\sum_{j=1}^{m} v_j \leq \sum_{i=1}^{m} b_i.$$

Also, any ℓ_j such that $m+1 \leq j \leq b$ satisfies $p_v \notin \ell_j$ so that $v_j \leq b_v \leq b(p)$ for any p, so if $b<v$ then $v_j \leq b_j$ for any $m+1 \leq j \leq b$. We therefore have the following inequalities:

$$\begin{aligned}
v_1 &\leq b_m \\
v_2 &\leq b_1 \\
v_3 &\leq b_2 \\
&\vdots \\
v_m &\leq b_{m-1} \\
v_{m+1} &\leq b_{m+1} \\
v_{m+2} &\leq b_{m+2} \\
&\vdots \\
v_b &\leq b_b.
\end{aligned}$$

† De Bruijn and Erdös (1948). Sometimes called the de Bruijn–Erdös and Hanani theorem because of Hanani (1955).

Thus

$$\sum_{i=1}^{v} b_i = \sum_{j=1}^{b} v_j \le \sum_{i=1}^{b} b_i$$

which is clearly impossible if $b < v$.

It also follows that if $v = b$, then $v_j = b_j$ for $m+1 \le j \le b = v$ and $b_i = v_{i+1}$, $1 \le i \le m-1$ and $b_m = v_1$. From the corollary to theorem 1.6.4 we have $\sum_{j=1}^{b} v_j(v_j - 1) = v(v-1)$ and the left hand side of this can now be rewritten as $\sum_{i=1}^{v} b_i(b_i - 1)$. But the right hand side is $b(b-1)$, so that the corollary to lemma 2.2.1 lets us conclude that all lines meet.

We consider two cases.

Case 1. Suppose that there are two lines ℓ and ℓ' which together contain all the points. Let $p = \ell \cap \ell'$. For $x \in \ell \backslash \{p\}$, and $x' \in l' \backslash \{p\}$, the line xx' can only have two points. Suppose both ℓ and ℓ' have third points y and y'. Then yy' is a 2-point line not meeting xx', so that this is not possible. So ℓ, say, has only two points and ℓ' has $v-1$ points. We call the space a *near-pencil.*

Case 2. Given any two lines ℓ and ℓ', suppose there is a point $p \notin \ell \cup \ell'$. We define a map f as follows. For each x in ℓ, $f: x \to px \cap \ell' = x'$. Note that if $x = \ell \cap \ell'$, $f: x \to x$. f is a map of ℓ onto ℓ' since, for any x' in ℓ', $f(x'p \cap \ell) = x'$. Also, f is 1–1 since, for $x \ne y$ in ℓ, the lines px and py are distinct and meet ℓ' in distinct points x' and y', say, using L2. So $f(x) = x'$ and $f(y) = y'$. Therefore ℓ and ℓ' have the same number of points. Since ℓ' can be thought of as an arbitrary line different from ℓ, we have proved that all lines have $|\ell|$ points. Let $|\ell| = k+1$. It follows immediately then, from the fact that all lines meet, that each point is on $k+1$ lines.

Clearly $k \ge 1$. In case $k = 1$ we have a triangle so there is no point p not in the union of two lines. In fact, the case $k = 1$ comes under case 1 above. So $k \ge 2$. This completes the proof. □

We have already seen an example of the type of space appearing in case 2. That is the Fano plane of figure 1.1.1.

2.3 Numerical properties

The axiom L2 allows us to produce some very nice results for linear spaces merely by counting. In this section we suppose that v (and therefore b) is finite.

Lemma 2.3.1. *For any fixed point* p_i, $\sum_{j=1}^{b} (v_j - 1) r_{ij} = v - 1$.

Proof. Any point is joined by a line to p_i because of L2. So, excluding p_i, we can count all points merely by counting the points of the lines on p_i. Each such line ℓ_j has $v_j - 1$ points if we exclude the point p_i. Again, by L2, this method counts each point (except p_i) precisely once.

So

$$v-1=\sum_{\ell_j \text{ on } p_i}(v_j-1).$$

But $p_i \in \ell_j$ if and only if $r_{ij}=1$. Hence

$$v-1=\sum_{j=1}^{b}(v_j-1)r_{ij}. \quad \square$$

Lemma 2.3.1 allows us to give now an alternative proof of the easier implication in the corollary to theorem 1.6.4.

Lemma 2.3.2. $\sum_{j=1}^{b} v_j(v_j-1)=v(v-1)$.

Proof. By lemma 2.3.1, $v-1=\sum_{j=1}^{b}(v_j-1)r_{ij}$ for a fixed point p_i. Now, summing over all points,

$$\sum_{i=1}^{v}(v-1)=\sum_{i=1}^{v}\sum_{j=1}^{b}(v_j-1)r_{ij}$$

$$v(v-1)=\sum_{j=1}^{b}(v_j-1)\sum_{i=1}^{v}r_{ij}$$

$$=\sum_{j=1}^{b}(v_j-1)v_j. \quad \square$$

Recall from section 1.5 that a (near-) linear space is point regular if $b_i=b_j$ for all $1\le i,j\le v$ or, equivalently, using the ordering of section 1.5, if $b_1=b_v$; and is line regular if $v_i=v_j$ for all $1\le i,j\le b$ or, equivalently, if $v_1=v_b$.

Lemma 2.3.3. *If a linear space is line regular then it is point regular.*

Proof. Fix a point, p_i. From lemma 2.3.1,

$$v-1=\sum_{j=1}^{b}(v_j-1)r_{ij}=kb_i$$

where $k+1$ is the line regularity.

$$\text{So}\quad b_i=\frac{v-1}{k}.$$

As this is independent of the choice of p_i, we have point regularity. $\square$

It is natural now to ask the converse of lemma 2.3.3. If a linear space is point regular, is it line regular? The answer is no. For instance, removing one point from the Fano plane of figure 1.1.1 and 'shortening' the appropriate lines does not affect the number of lines per point. Removing the point 0, say, leads to the linear space of figure 2.3.1 which has point regularity 3, but which does not have line regularity. (This is sometimes called the 'punctured' Fano plane.)

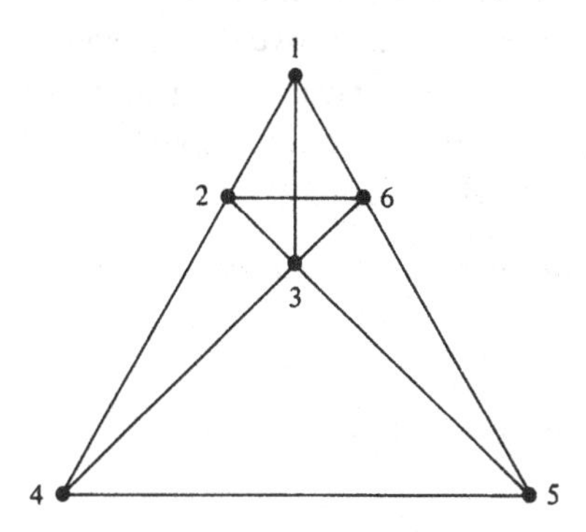

Figure 2.3.1.

Lemma 2.3.4. *Let a linear space have line and point regularity $k+1$, $k \geq 1$. Then all lines meet and $b = v = k^2 + k + 1$. (The space is then the type appearing in the latter part of (ii) in the de Bruijn–Erdös theorem, theorem 2.2.2.)*

Proof. Let ℓ and $\hbar$ be any two lines, and let $p \in \hbar$. If $p \in \ell$, we are done. Otherwise, each line on p meets ℓ as there are $k+1$ lines on p and $k+1$ points on ℓ. In particular, $\hbar$ meets ℓ.

To determine v, fix a point p. By lemma 2.3.3

$$b(p) = \frac{v-1}{k} = k+1.$$

So $v - 1 = k^2 + k$ or $v = k^2 + k + 1$.

Now $v(k+1)$ is the number of points multiplied by the number of lines per point. Each line is repeated $k+1$ times in this calculation, so that

$$b = \frac{v(k+1)}{k+1} = v. \quad \square$$

If a linear space S is line regular of regularity $k+1$, $k \geq 1$ and if $b_1 \leq k+1$, what can we say? By lemma 2.3.3, S is point regular of regularity $(v-1)/k \leq k+1$. If there is a point p not on some line ℓ, then $b(p) \geq k+1$ implies $b(p) = k+1$. And this is true for *each* point not on ℓ. Moreover, for each point q of ℓ, we can find lines not on it and argue in the same way. So, in fact, if there are at least two lines, S has point regularity $k+1$. If there is only one line, clearly S has point regularity 1. So we have proved the next lemma.

Lemma 2.3.5. If a linear space S with $b \geq 1$ has line regularity $k+1$, $k \geq 1$ and if $b_1 \leq k+1$, then S has point regularity either 1 or $k+1$.

Theorem 2.2.2 and lemma 2.3.4 present conditions under which all lines meet. In fact, given that all lines meet, we can prove that, as soon as $b > 1$, we have one of the two possible types of space mentioned in part (ii) of theorem 2.2.2. The proof of this can be culled immediately from that of the theorem. We summarize the result as follows.

Lemma 2.3.6. If all lines of a linear space S meet then either

(i) *S is trivial, or*
(ii) *S is a near-pencil, or*
(iii) *S has point and line regularity $k+1$, for some $k \geq 2$ and $v = b = k^2 + k + 1$.*

2.4 The exchange property

In section 1.4 we introduced concepts of independence, basis and dimension. We saw that spaces may have bases of different orders and promised in chapter 2 to consider a property which would eliminate this possibility. We do this in this section. The theory we develop is equally valid in near-linear spaces.

A linear space has the *exchange property* if, for all points x and y and sets of points X,

$$x \notin \langle X \rangle \text{ and } x \in \langle X \cup \{y\} \rangle \text{ imply } y \in \langle X \cup \{x\} \rangle.$$

In other words, one is able to *exchange* x and y. We emphasize that this must be true for all possible choices of x, y and X.

As an example, consider the linear space of figure 2.4.1.

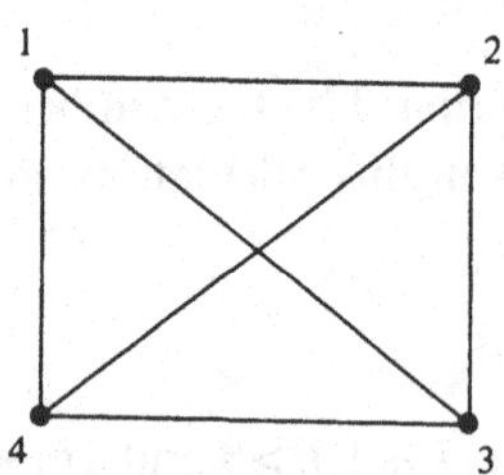

Figure 2.4.1.

Choosing $x = 1$, $y = 2$ and then considering choices for X, trying to satisfy $x \notin \langle X \rangle$ while $x \in \langle X \cup \{y\} \rangle$ is impossible. So there is nothing to prove in the implication. If $x = y = 1$ and $X = \{3,4\}$ then everything is satisfied. The reader will see that, after checking all possibilities, this space does indeed have the exchange property.

The space of figure 1.1.1 also has the exchange property. For instance, if $x=1$, $y=2$ and $X=\{4\}$ we see that $x\notin\langle X\rangle=\{4\}$, $x\in\langle X\cup\{y\}\rangle=\{1,2,4\}$ and also $y\in\langle X\cup\{x\}\rangle=\{1,2,4\}$. We urge the reader to test the other possibilities.

Of course, not all linear spaces should have the exchange property. Consider the space of figure 2.1.1.

$6\notin\langle\{2,7\}\rangle$, $6\in\langle\{2,7\}\cup\{3\}\rangle$ while
$3\notin\langle\{2,7\}\cup\{6\}\rangle$.

(Note that for a linear space not to have the exchange property, it is enough to find one choice of x, y and X for which our 'switching' does not work.)

As mentioned, spaces with the exchange property behave nicely. The reader should check the space of figure 2.1.1 with regard to each of the lemmas in this section.

Lemma 2.4.1. *Let S be a linear space with exchange property. If $X=\{x_1,x_2,\ldots,x_n\}$ is an independent set and $x_{n+1}\notin\langle X\rangle$, then $\{x_1, x_2,\ldots,x_n, x_{n+1}\}$ is independent.*

Proof. We must prove that no $x_i\in\langle\{x_1,x_2,\ldots,x_{n+1}\}\backslash\{x_i\}\rangle$. If $x_i=x_{n+1}$ this follows from our hypothesis. Suppose $x_i\neq x_{n+1}$, and $x_i\in\langle\{x_1,x_2,\ldots,x_{n+1}\}\backslash\{x_i\}\rangle=\langle\{x_1,x_2,\ldots,x_n\}\backslash\{x_i\}\cup\{x_{n+1}\}\rangle$. But $x_i\notin\{x_1,x_2,\ldots,x_n\}\backslash\{x_i\}$ as X is independent so, by the exchange property, $x_{n+1}\in\langle\{x_1,x_2,\ldots,x_n\}\backslash\{x_i\}\cup\{x_i\}\rangle=\langle X\rangle$, a contradiction. □

Lemma 2.4.2. *Let S be a linear space with the exchange property. If $q\in\langle X\cup\{p\}\rangle$ but $q\notin\langle X\rangle$ then $\langle X\cup\{p\}\rangle=\langle X\cup\{q\}\rangle$.*

Proof. The properties of closure imply $\langle X\cup\{q\}\rangle\subseteq\langle X\cup\{p\}\rangle$. But, since S has the exchange property, $p\in\langle X\cup\{q\}\rangle$ and so $\langle X\cup\{p\}\rangle\subseteq\langle X\cup\{q\}\rangle$ also. □

A subspace V *covers* a subspace W if $W\subsetneq V$ and if $W\subsetneq H\subseteq V$ for any subspace H implies $H=V$; that is, if V and W are different and there is no subspace between them. We say also that W *is covered by* V.

Notation: $W\prec V$ or $V\succ W$.

Lemma 2.4.3. *Let S be any linear space. If V and W are subspaces such that $W\prec V$ then $V=\langle W\cup\{p\}\rangle$ for some point p.*

Proof. Choose any $p\in V\backslash W$. Then $W\subsetneq\langle W\cup\{p\}\rangle\subseteq V$ so that $V=\langle W\cup\{p\}\rangle$. □

Lemma 2.4.4. *Let S be a linear space with the exchange property. If $p\notin\langle X\rangle$, then $\langle X\rangle\prec\langle X\cup\{p\}\rangle$.*

Proof. Suppose $\langle X\rangle \subsetneq V \subseteq \langle X\cup\{p\}\rangle$ where V is a subspace. Then choosing $q\in V\backslash\langle X\rangle$ it follows that $\langle X\cup\{q\}\rangle \subseteq V \subseteq \langle X\cup\{p\}\rangle$. By lemma 2.4.2, $\langle X\cup\{q\}\rangle = \langle X\cup\{p\}\rangle$, and hence $V=\langle X\cup\{p\}\rangle$. □

Lemma 2.4.5. *Let S be a finite linear space with the exchange property. Then any two bases have the same number of elements.*

Proof. The proof we give here is rather involved and resorts to saying after a few steps 'and so on'; but it is descriptive. We suggest that the reader attempt a neater proof, say by induction.

Let $\{x_1,x_2,\ldots,x_n\}$ and $\{y_1,y_2,\ldots,y_m\}$ be bases with $n\le m$. It cannot be the case that each $y_i\in\langle\{x_2,x_3,\ldots,x_n\}\rangle$ as otherwise $S=\langle\{y_1,y_2,\ldots,y_m\}\rangle\subseteq\langle\{x_2,x_3,\ldots,x_n\}\rangle\subsetneq S$.

Suppose $y_1\notin\langle\{x_2,x_3,\ldots,x_n\}\rangle$. Since $y_1\in\langle\{x_1,x_2,\ldots,x_n\}\rangle$, the exchange property implies $x_1\in\langle\{y_1,x_2,\ldots,x_n\}\rangle$. Furthermore, $y_1\notin\langle\{x_2,x_3,\ldots,x_n\}\rangle$ and $x_2\notin\langle\{x_3,x_4,\ldots,x_n\}\rangle$ implies $x_2\notin\langle\{y_1,x_3,\ldots,x_n\}\rangle$, using the exchange property. So $\langle\{y_1,x_3,\ldots,x_n\}\rangle\subsetneq\langle\{y_1,x_2,\ldots,x_n\}\rangle=S$. If each y_i is in $\langle\{y_1,x_3,\ldots,x_n\}\rangle$, we get $S\subseteq\langle\{y_1,x_3,\ldots,x_n\}\rangle\subsetneq S$, clearly a contradiction.

Suppose $y_2\notin\langle\{y_1,x_3,\ldots,x_n\}\rangle$. Then $y_2\in\langle\{y_1,x_2,\ldots,x_n\}\rangle=S$ so that, by the exchange property, $x_2\in\langle\{y_1,y_2,x_3,\ldots,x_n\}\rangle=S$.

Continuing in this way we get, eventually, $S=\langle\{x_1,x_2,\ldots,x_n\}\rangle\subseteq\langle\{y_1,y_2,\ldots,y_n\}\rangle\subseteq\langle\{y_1,y_2,\ldots,y_m\}\rangle=S$ so that $n=m$. □

Any subspace V of a near-linear space can be treated as a near-linear space in its own right, and thus has a dimension which we denote by dim V.

Lemma 2.4.6. *Let S be a linear space with the exchange property. If* $p\notin\langle X\rangle$ *and* dim $\langle X\rangle = n$ *then* dim $\langle X\cup\{p\}\rangle = n+1$.

Proof. By lemma 2.4.5, any basis of $\langle X\rangle$ has $n+1$ elements. Let $\{x_1,x_2,\ldots,x_{n+1}\}$ be such a basis. By lemma 2.4.1 $\{x_1,x_2,\ldots,x_{n+1},p\}$ is independent and clearly it generates its closure, $\langle X\cup\{p\}\rangle$. As any two bases of $\langle X\cup\{p\}\rangle$ have the same number of elements, this number is $n+2$. So dim $\langle X\cup\{p\}\rangle = n+1$. □

2.5 Hyperplanes

A *hyperplane* of a linear space S is a subspace which is covered by S. Equivalently, it is a maximal proper subspace.

Example 2.5.1. In real 5-space, the hyperplanes are the 4-spaces.

Example 2.5.2. In the linear spaces of figures 1.1.1 and 2.1.2, the hyperplanes are the lines.

Example 2.5.3. The hyperplanes of a line are the points; the only hyperplane of a point is the empty set. The empty set itself has no hyperplanes.

It would seem natural that the dimension of a hyperplane be 1 less than the dimension of the space. This in fact is not true, as can be seen by once again examining the linear space of figure 2.1.1 where the set $\{3, 5, 6, 7, 8\}$ forms a hyperplane with dimension 2 while S itself has dimension 2. However, in the presence of the exchange property, this can no longer occur.

Lemma 2.5.1. *Let S be a finite linear space with the exchange property. If H is a hyperplane of S then* $\dim H = \dim S - 1$.

Proof. As S covers H, $S = \langle H \cup \{p\} \rangle$ for any point $p \in S \backslash H$. By lemma 2.4.6, $\dim \langle H \cup \{p\} \rangle = \dim H + 1$. □

A *projective plane* is a linear space in which

1. any two lines meet,
2. there exists a set of four points no three of which lie on a single line.

The second condition implies that there are at least two lines. Near-pencils, however, are excluded, even though they satisfy axiom 1. We have seen only two examples of projective planes so far, but we shall be seeing more of them in chapter 3. The two examples are the Fano plane (figure 1.1.1) and the extended real plane (example 2.1.3). We leave it to the reader to verify that they are indeed projective planes.

Lemma 2.5.2. *The hyperplanes of a projective plane are precisely the lines.*

Proof. It suffices to show that if ℓ is any line and if $x \notin \ell$ then $\langle \ell \cup \{x\} \rangle = S$. To do this we show that $S \subseteq \langle \ell \cup \{x\} \rangle$ or, equivalently, for any point p of S, $p \in \langle \ell \cup \{x\} \rangle$. If $p = x$ or $p \in \ell$, we are finished. Otherwise px is a line and it meets ℓ in a unique point q, say. But q and x are both in $\langle \ell \cup \{x\} \rangle$ and, since $\langle \ell \cup \{x\} \rangle$ is a subspace, $p \in qx \subseteq \langle \ell \cup \{x\} \rangle$. □

If a projective plane had the exchange property, we could say immediately from lemma 2.5.1 that it had dimension 2.

Lemma 2.5.3. *A projective plane has the exchange property.*

Proof. We must prove that, for any points x, y and any set X, $x \notin \langle X \rangle$ and $x \in \langle X \cup \{y\} \rangle$ imply $y \in \langle X \cup \{x\} \rangle$. There are precisely three possibilities for $\langle X \rangle$. It is $\varnothing$, a point or a line.

If $\langle X \rangle = \varnothing$, then $x \in \langle X \cup \{y\} \rangle$ implies $x = y$ and the rest is trivial. If $\langle X \rangle = p$, a point, then $x \in \langle X \cup \{y\} \rangle$ means $x \in py$ so $y \in px = \langle X \cup \{x\} \rangle$. If $\langle X \rangle = \ell$, a line, then $x \notin \ell$ implies $\langle \ell \cup \{x\} \rangle = S$ by lemma 2.5.2. So it is trivial that $y \in \langle X \cup \{x\} \rangle$. □

We have proved something slightly stronger:

Lemma 2.5.4. *Any linear space in which the lines are precisely the hyperplanes satisfies the exchange property.*

Lemma 2.5.5. *A projective plane has dimension* 2.

A *projective hyperplane* H is a proper subspace of a linear space S such that each line of S has a point in H.

In the linear spaces of figures 1.1.1 and 2.3.1 any 3-point line is a projective hyperplane.

The space of figure 1.7.1 can be transformed into a linear space in which the only projective hyperplane is the set $\{b, d, g\}$.

Lemma 2.5.6. *Any projective hyperplane is a hyperplane.*

Proof. Let H be a projective hyperplane of S and let $p \in S \backslash H$. If q is any point different from p and not in H, then by assumption the line pq meets H in a point r, say. But r and p in $\langle H \cup \{p\} \rangle$ implies $q \in rp \subseteq \langle H \cup \{p\} \rangle$. Hence $S \subseteq \langle H \cup \{p\} \rangle$ and we are done. □

2.6 Linear functions

It is not difficult to find examples of linear functions which map sets of a near-linear space which are *not* lines to lines in some other near-linear space. However, in the next lemma we show that if the domain of f is a linear space then the image of f must also be a linear space.

Lemma 2.6.1. *Let S be a linear space, S′ a near-linear space and f a linear function from S onto S′. Then S′ is a linear space.*

Proof. Let p and q be distinct points of S'. Since f is onto there are distinct points x and y of S such that $f(x) = p$ and $f(y) = q$. Since f is linear $f(xy)$ must be a line on p and q. □

Recalling the definition of isomorphic near-linear spaces (section 1.7) we saw that to get nice results we needed to assume f^{-1} linear. (See exercise 26 of section 1.8.) When dealing with linear spaces S and S', as soon as f is a 1–1 linear function from S onto S', it follows that f^{-1} is linear, as we show in the next lemma.

Lemma 2.6.2. *Let S and S′ be linear spaces and f a* 1–1 *linear function from S onto S′. Then f^{-1} is linear.*

Proof. Let p and q be distinct points of S'. Since f is 1–1 and onto, there are unique distinct points x and y of S such that $f(x) = p$ and $f(y) = q$. Since f is linear $f(xy) = pq$ and hence $f^{-1}(pq) = xy$. □

For the remainder of this section, we examine how f and f^{-1} operate on subspaces.

Let S and S' be the linear spaces of figure 2.6.1.

Define a linear function f by $f: 1 \to d,\ 2 \to c,\ 3 \to b,\ 4 \to a,\ 5 \to d$. (The reader should check that this is indeed linear.)

The subspaces of S are $\varnothing$, each point, each line and S itself. What does f do to these subspaces? $f(\varnothing) = \varnothing$; $f(x)$ is a point of S' for any point x of S; $f(\ell)$ is a line of S' for each line of ℓ, because f is linear. $f(S) = \{a, b, c, d\}$ is in fact a subspace of S'. We shall see in the next lemma that the image of a subspace under a linear function is always a subspace.

Lemma 2.6.3. *Let V be a subspace of S and let f be a linear function from S into S'. Then $f(V)$ is a subspace of S'.*

Proof. We must show that if p and q are distinct points of $f(V)$, then the line pq is a subset of $f(V)$. Since $p, q \in f(V)$ there are points x and y in V such that $p = f(x)$ and $q = f(y)$. Also, $x \neq y$ as f is a function. Then f maps the line xy to a line $f(xy)$ as f is linear. But the line $f(xy)$ contains the points p and q so that $f(xy) = pq$. So $pq \subseteq f(V)$. □

Lemma 2.6.4. *If f is a linear function from S to S', and if $x \neq y$ are points of S such that $f(x) \neq f(y)$ then $f(xy) = f(x)f(y)$.*

Proof. This is clear, and in fact we used it in the proof of the last lemma. □

Looking again at lemma 2.6.3, it is natural to ask if a subspace of S' always comes from a subspace of S. That is, if V is a subspace of S' is $f^{-1}(V)$ a subspace of S? (We emphasize that we are not assuming that f is 1–1. $f^{-1}(V)$ simply means the set of points of S which are mapped to points of V.) The answer to the question is no. To see this, consider the linear spaces of figure 2.6.1 with the function f given there. Let $V = \{a, d\}$, a line, then $f^{-1}(V) = \{1, 4, 5\}$ is not a subspace of S. If f is 1–1, then we can go backwards.

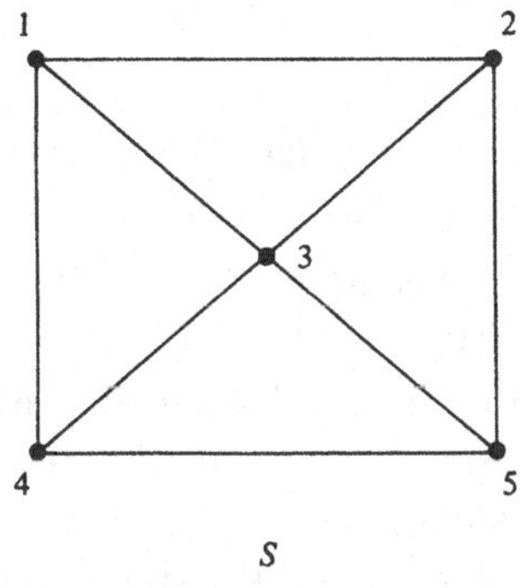

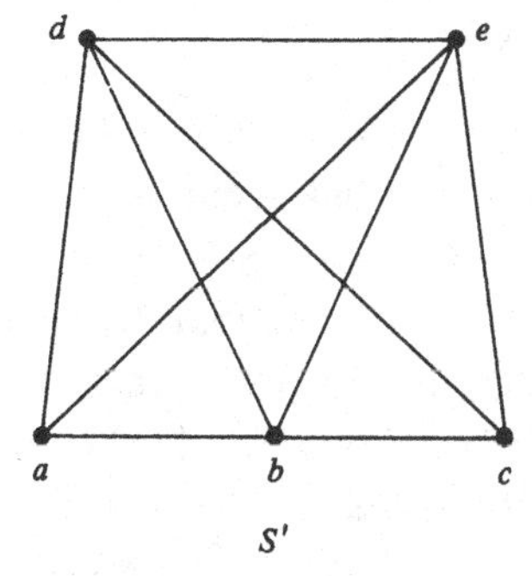

Figure 2.6.1.

Lemma 2.6.5. *Let V be a subspace of S' and f a 1–1 linear function from S into S'; then $f^{-1}(V)$ is a subspace of S.*

Proof. Let x and y be distinct points of $f^{-1}(V)$. We want to show that $xy \subseteq f^{-1}(V)$. Let p and q be points of V such that $f(x)=p$ and $f(y)=q$. Since f is 1–1, $p \neq q$, and so the line $pq \subseteq V$. But f is linear and $f(xy)$ is a line. In fact, by lemma 2.6.4 it is the line $f(x)f(y)=pq$. So $xy \subseteq f^{-1}(pq) \subseteq f^{-1}(V)$. □

Lemma 2.6.6. *Let f be a 1–1 linear function from S into S'. If V is a subspace of S then $f^{-1}f(V)=V$. If V' is a subspace of S', then $ff^{-1}(V') \subseteq V'$. If f is an onto function also, then $ff^{-1}(V')=V'$.*

Proof. Since f^{-1} is not necessarily defined on the whole of S', we can only conclude that $ff^{-1}(V') \subseteq V'$ in general. But, if f is onto, f^{-1} is a function from S' to S and we get equality. (See exercise 55 of section 2.7.) □

Lemma 2.6.7. *Let X be a subset of S and f a 1–1 linear function from S into S'. Then, for any subspace V on X, $f(V)$ is a subspace of S' on $f(X)$ and, for any subspace V' on $f(X)$, $f^{-1}(V')$ is a subspace of S on X. Moreover, if f is onto, there is a 1–1 correspondence between subspaces of S on X and subspaces of S' on $f(X)$.*

Proof. This follows quickly from lemmas 2.6.3, 2.6.5 and 2.6.6. □

Lemma 2.6.8. *Let X be a subset of S and f a 1–1 linear function from S into S'. Then $\langle f(X)\rangle = f(\langle X\rangle)$.*

Proof. By lemma 1.4.1, $\langle f(X)\rangle = \bigcap\{V' \mid V' \text{ a subspace of } f(S) \text{ on } f(X)\}$.

By lemma 2.6.7, any subspace V of S on X gives rise to a subspace $f(V)$ on $f(X)$. Conversely, any subspace V' of S' on $f(X)$ gives rise to a subspace V of S on X. Thus $f(\bigcap\{V \mid V \text{ a subspace of } S \text{ on } X\}) = \bigcap\{V' \mid V' \text{ a subspace of } f(S) \text{ on } f(X)\}$ or, equivalently, $f(\langle X\rangle) = \langle f(X)\rangle$. □

Lemma 2.6.9. *Let f be a 1–1 linear function from S into S'. If V is a subspace of S, then* $\dim V \geq \dim f(V)$.

Proof. Choose a basis B of V. Then, by lemma 2.6.8, $V=\langle B\rangle$ implies $f(V)=\langle f(B)\rangle$ so that $f(B)$ generates $f(V)$.

Suppose for some $x \in B$, $f(x) \in \langle f(B)\backslash\{f(x)\}\rangle$. By lemma 2.6.8, this implies $f(x) \in f(\langle B\backslash\{x\}\rangle)$ and so, since f is 1–1, $x \in \langle B\backslash\{x\}\rangle$, contradicting the fact that B is independent.

So any basis of V produces a basis of $f(V)$, and the above inequality follows because of the definition of dimension. □

Lemma 2.6.10. *Let f be an isomorphism of S onto S'. If V' is a subspace of S', then* $\dim V' \geq \dim f^{-1}(V')$.

Proof. This follows immediately from lemmas 2.6.2 and 2.6.9. □

Combining the last two lemmas gives us something much stronger.

Theorem 2.6.11. *Let f be a* 1–1 *linear function from S into S'. Then for any subspace V of S,* $\dim V = \dim f(V)$.

Proof. We may consider f to be an isomorphism from S onto the linear space $f(S) \subseteq S'$. Thus $f(V)$ is a subspace of $f(S)$ (see exercise 2.7.34). By lemma 2.6.9, $\dim V \geq \dim f(V)$ and, by lemma 2.6.10, since f^{-1} is an isomorphism of $f(S)$ onto S, $\dim f(V) \geq \dim f^{-1}(f(V))$ which is just V by lemma 2.6.6. Then $\dim V \geq \dim f(V) \geq \dim V$ yields inequality. □

Corollary. *If f is an isomorphism from S onto S' and V' is a subspace of S', then* $\dim V' = \dim f^{-1}(V')$.

2.7 Exercises

1. Find all (non-isomorphic) linear spaces on six points. (There are ten.)
2. Show that the dual of a linear space is not necessarily linear.
3. When is the dual of a linear space linear?
4. Is a restriction of a linear space always a linear space?
5. Find a linear space which has infinite dimension.

*6. Show that it is possible to have two non-isomorphic linear spaces with the same parameters v, b, v_i, b_i. (See exercise 32 of section 1.8 and the last paragraph of section 3.10.).

7. Let S have six points, seven lines and no 4-point lines. Draw S.
8. Let S satisfy $(b-v)^2 \leq v$. (Such a linear space is called *restricted*.) Find all such spaces with $v = 6$.
9. Construct a non-trivial $S(3, k, v)$.
10. Show that the Fano plane is the unique $S(2, 3, 7)$.
11. Find the parallel classes (see section 2.1) for the linear space of figure 2.1.2.
12. An *equivalence relation* R on a set S is a subset of $S \times S$. We say a is related to b, and write aRb if $(a, b) \in R$. The following properties must also hold:

    aRa (reflexive property),

    aRb implies bRa (symmetric property),

    aRb and bRc imply aRc (transitive property), for all a, b, c in S.

    Show that 'parallelism' (i.e. $\ell \| k$ if and only if $\ell = k$ or ℓ misses k) is an equivalence relation on the lines of $\mathbb{R}^2$.
13. If $v_i = k$, $1 \leq i \leq b$, and $b_i \leq k+1$, $1 \leq i \leq v$, show that parallelism is an equivalence relation.

14. Show that the de Bruijn–Erdös theorem is not necessarily true in near-linear spaces.

15. If $b > 1$, is $b \geq v$ enough to ensure that a near-linear space is linear?

16. Find an example of a space of the second type described in the de Bruijn–Erdös theorem, with $k = 3$.

17. Prove lemma 2.3.6.

18. Find an example different from that of figure 2.3.1 which is point regular but not line regular.

19. Prove that all punctured Fano planes are isomorphic.

20. Let S be the restriction of $\mathbb{R}^2$ to the points on the circumference of a given circle. Is S point regular? Is it line regular?

21. Show that if $v = 5n + 3$ there is no linear space with only 5-lines or 6-lines. (Hint: use the equation of lemma 2.3.2.)

22. Let S have $v = n^2 + n$ points, each line an n-line or an $(n + 1)$-line. Suppose there are precisely two $(n + 1)$-lines and that these are disjoint. Describe S.

23. If all lines of S are n-lines, $v < n^2$ and there are two parallel n-lines, show that S does not exist.

24. If S has $v = n^2 + n + 1$ points, each line is an n-line or an $(n + 1)$-line and there is at most one $(n + 1)$-line, what can you say about S?

25. Is it possible to have a linear space with only 3-point lines and $v = 8$?

26. Suppose S is a linear space with $v \geq n^2$ and $2n^2 + 3n \geq 35$, n a positive integer. Show that $v \geq 16$. (Hint: solve a quadratic equation.)

27. Draw a linear space associated with the matrix

$$M = \begin{pmatrix} 1 & 0 & 0 & 1 & 0 & 1 \\ 1 & 0 & 1 & 0 & 1 & 0 \\ 1 & 1 & 0 & 0 & 0 & 0 \\ 0 & 1 & 1 & 0 & 0 & 1 \\ 0 & 1 & 0 & 1 & 1 & 0 \end{pmatrix}$$

28. Compute MM^t for the matrix M of exercise 27.

29. Show that any finite dependent set of points X of a linear space contains an independent set Y such that $\langle Y \rangle = \langle X \rangle$.

30. Find a linear space different from that of figure 2.1.1 which does not have the exchange property.

31. Find an example, using lemma 2.5.4, of a linear space, not a projective plane, which has the exchange property.

32. If S has the exchange property and V and W are subspaces such that $W \nsubseteq V$ and dim $V = m$ is finite, show that dim $\langle V \cup W \rangle \geq m + 1$.

33. Use the exchange property to prove that if U and V are subspaces of a linear space such that U, $V \succ W$ and if $U \neq V$, then $\langle U \cup V \rangle \succ U$ and $\langle U \cup V \rangle \succ V$.

34. If U is a subspace of V and V is a subspace of W, in a linear space S, show that U is a subspace of W.

35. If V is a restriction of S and U is a subspace of V, is U a subspace of S?

36. Does the linear space of figure 2.1.1 have any projective hyperplanes?

37. If H is a projective hyperplane of S and V is a proper subspace of S not contained in H, prove that $H \cap V$ is a projective hyperplane of V.

38. Find a linear space with at least two hyperplanes of dimension greater than or equal to 2.

39. Is any hyperplane of a linear space a projective hyperplane?

40. Let $f((x, y)) = (x + a, y + b)$ for fixed constants a and b be a function defined on $\mathbb{R}^2$. Is f a linear function? (f is called a *translation*.)

41. The function $f((x, y, z)) = (x, z - y, x + z)$ is defined on $\mathbb{R}^3$. Prove that it is 1–1 and onto. Is it linear?

42. Prove that the set of all collineations of a linear space forms a group. (See lemma 1.7.4.)

43. Find all linear spaces for which every subset of points is a subspace.

44. Find all collineations of the complete graph K_n. (See exercise 11 of section 1.8.)

***45.** Prove that if any possible permutation of the points of a linear space S is a collineation of S, then S is a complete graph.

46. Explain why there are 168 collineations of the Fano plane.

47. Compute the number of collineations of the linear space of figure 2.1.2.

48. Let $S = \mathbb{R}^2$ and define f as follows. For distinct fixed points p and q, $f(p) = q$ and $f(q) = p$ and otherwise $f(x) = x$. Is f a collineation?

49. Show that the conclusion of lemma 2.6.1 does not necessarily hold if f is not onto.

50. Prove lemma 2.5.5 directly. That is, show there is a basis with three elements.

***51.** Prove lemma 2.4.5 by induction.

52. Two sets have the same number of elements if there is a 1–1 onto function between them. Show that the set of integers and the set of positive integers have the same number of elements.

53. A *closure space* is a set S with a function ϕ from $\mathscr{P}(S)$ (the set of all subsets of S) back to $\mathscr{P}(S)$ such that for X, $Y \in \mathscr{P}(S)$,

$X \subseteq \phi(X)$,
$\phi(X) = \phi(\phi(X))$,

$X \subseteq Y$ implies $\phi(X) \subseteq \phi(Y)$. (Compare with section 1.4.) Construct two different closure spaces from the set $S = \{a, b, c, d\}$.

54. Show that for any linear space S, the function ϕ which maps a set of points X of S to its closure $\langle X \rangle$ turns S into a closure space.

55. Find an example exhibiting the case $ff^{-1}(V') \subsetneq V'$ of lemma 2.6.6.

3

Projective spaces

...and as they journeyed on, a whisper thrilled along the superficies in isochronous waves of sound, 'Yes! We shall at length meet if continually produced!'

Lewis Carroll *The Dynamics of a Particle*

For most of this chapter (sections 3.1–3.8) we shall be dealing with the notion of projective plane, before introducing the more general notion of projective space.

3.1 Projective planes

Recall (section 2.5) that a *projective plane* is a linear space in which

PP1 any two lines meet,

PP2 there exists a set of four points no three of which are collinear.

It follows from L2 that any two lines meet in a unique point.

So far we have seen two examples: the Fano plane (figure 1.1.1) and the extended real plane (example 2.1.3). We present one other example in this section, but will see many more examples later in the chapter, along with examples of more general projective spaces.

The linear space of figure 3.1.1 is a projective plane with thirteen points $\{1,2,\ldots,13\}$ and thirteen lines $\{\{1,2,3,11\}$, $\{4,5,6,11\}$, $\{7,8,9,11\}$, $\{1,4,7,13\}$, $\{2,5,8,13\}$, $\{3,6,9,13\}$, $\{1,5,9,12\}$, $\{2,6,7,12\}$, $\{3,4,8,12\}$, $\{1,6,8,10\}$, $\{2,4,9,10\}$, $\{3,5,7,10\}$, $\{10,11,12,13\}\}$.

We also recall two lemmas from section 2.5 (lemmas 2.5.3 and 2.5.5) reproducing them here as lemmas 3.1.1 and 3.1.2.

Lemma 3.1.1. *A projective plane has the exchange property.*

Lemma 3.1.2. *A projective plane has dimension* 2.

Lemma 3.1.2 gives us the 'right' to call a projective plane a plane.

Recalling the labelling of points and lines from chapter 2, we are able to prove the next lemma.

Lemma 3.1.3. *If a linear space contains four points, no three of which are collinear, and has the property that $c_{ij} = b_i$ for every p_i and ℓ_j such that $r_{ij} = 0$, then it is a projective plane.*

Proof. The fact that $c_{ij} = b_i$ whenever $r_{ij} = 0$ ensures that all lines meet. □

If $\mathscr{S}$ is any statement about points and lines, the *dual statement* of $\mathscr{S}$ is the statement formed from $\mathscr{S}$ by interchanging the words 'line' and 'point', and 'contains' and 'is contained in' (or any given equivalents).

Example 3.1.1. If $\mathscr{S}$ is the statement 'each line has seven points', the dual of $\mathscr{S}$ would be 'each point is on seven lines'.

Lemma 3.1.4. *In a projective plane, the dual statements of L1, L2, PP1 and PP2 hold.*

Proof. The dual of PP1 is the statement 'any two points are on a common line', and this is always true in a projective plane. The dual of PP2 states that 'there exists a set of four lines no three of which are on a common

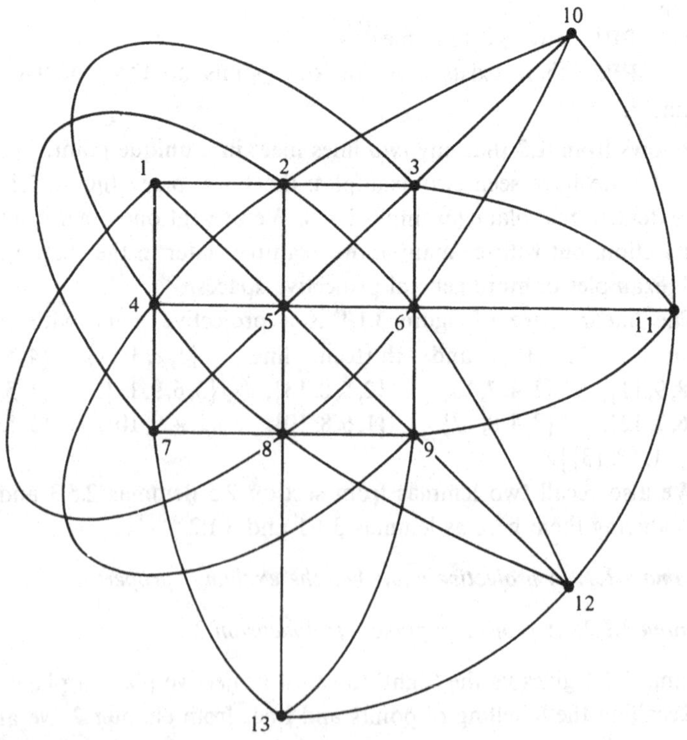

Figure 3.1.1.

point'. By PP2 there are points 1, 2, 3, 4 no three collinear. Hence the lines 12, 23, 34 and 14 are distinct and no three are on a common point.

We leave the dual statements of L1 and L2 as an exercise. □

Lemma 3.1.5. *If $\mathscr{S}$ is any statement that holds in a projective plane, then the dual statement of $\mathscr{S}$ also holds.*

Proof. Any statement $\mathscr{S}$ which is valid in a projective plane can be proved from L1, L2, PP1 and PP2. Since the duals of these latter statements are true, the dual of $\mathscr{S}$ is. (One could repeat the proof of $\mathscr{S}$ replacing all statements by their duals.) □

Lemma 3.1.5 cuts our work-load in half: we need only prove 50% of the theorems about projective planes. In fact, projective planes have many beautiful properties, as we shall see in the following sections.

3.2 Finite projective planes

We assume in this section that Π is a projective plane with a finite number v of points and a finite number b of lines.

Lemma 3.2.1. *Π has point and line regularity $k+1$, say, $k \geq 2$, and $v = b = k^2+k+1$.*

Proof. This follows immediately from lemma 2.3.6 because of PP2. □

We call k the *order* of the projective plane.

Notice that the dual of lemma 3.2.1 is itself.

We have now seen examples of projective planes of orders 2 and 3. It seems reasonable to assume that one could construct a projective plane of order k for any $k \geq 2$, but this is in fact not the case. It is known that there are unique projective planes of orders 2, 3, 4, 5, 7 and 8. There is no projective plane of order 6. There are precisely four non-isomorphic projective planes of order 9.[†] The case of order 10 has recently been settled in the negative by a lengthy computer search.[‡]

Lemma 3.2.2. *There is a unique projective plane of order* 2.

Proof. (See figure 1.1.1.) As the point 1 is on three lines, we may assume without loss of generality that these lines are $\{1,2,4\}$, $\{1,3,0\}$ and $\{1,6,5\}$. The point 2 is also on three lines, one of which is $\{1,2,4\}$. Again without loss of generality, we may assume that the other lines are $\{2,3,5\}$ and $\{2,0,6\}$. Now 4 and 0 must be on a line and the third point of this line can only be the point 5 because of L2. Similarly the line on 4 and 3 contains 6. We now have seven lines and it can easily be checked that we have

[†] See C. W. H. Lam, G. Kolesova and L. Thiel (1991).
[‡] See C. W. H. Lam, L. Thiel and S. Swiercz (1989).

a projective plane of order 2. Since this is the only possible construction on seven points, apart from relabelling the points, it is unique. □

3.3 Embedding a near-linear space in a projective plane

An *embedding* of a near-linear space $S=(P,L)$ into a near-linear space $S'=(P',L')$ is a function f mapping P into P' and L into L' which is one-to-one on both points and lines, and such that $p\in\ell$ if and only if $f(p)\in f(\ell)$.

Example 3.3.1. Let S and S' be the linear spaces of figures 3.3.1 (*a*) and (*b*) respectively.

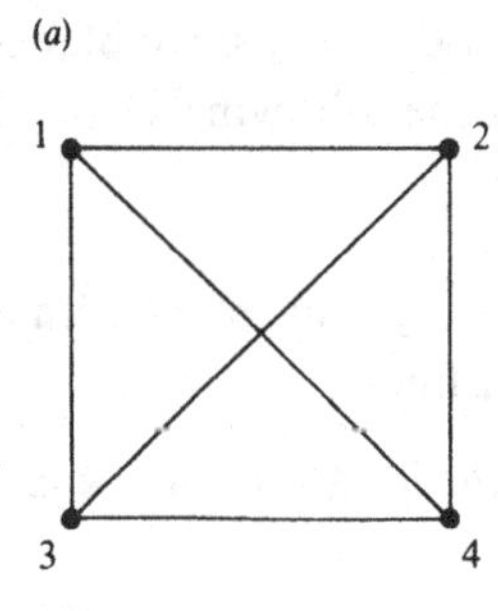

Figure 3.3.1.

The map f: $1\to b$, $2\to g$, $3\to c$, $4\to e$ induces an embedding from S into S' such that $\{1,2\}\to\{b,g,d\}$, $\{2,4\}\to\{a,g,e\}$, $\{3,4\}\to\{c,d,e\}$, $\{1,3\}\to\{a,b,c\}$, $\{1,4\}\to\{b,h,e\}$ and $\{2,3\}\to\{c,h,g\}$. Notice that f is onto neither on the points nor the lines, and it is certainly not a linear function.

Marshall Hall (1943) showed that any near-linear space in which one can find four points, no three collinear, is embeddable in a projective plane. We prove this again here.

Theorem 3.3.1. *A near-linear space S in which there is a set of four points, no three collinear, can be embedded in a projective plane.*

Proof. We build the projective plane around the near-linear space by adding points and lines wherever necessary. PP2 holds by assumption, and we must make sure that any two points are on a unique line, two lines intersect in a unique point, and lines have at least two points.

Consider all pairs of points of S not already on a line. Then add all the 2-point lines necessary to make S into a linear space. Now consider lines which do not meet. Add a new point to any such pair of lines. Now we have again points not joined by a line, so add 2-point lines.

The method is now clear. Continuing in this way, and taking the union of the near-linear spaces obtained at each step, we get a projective plane, as it is now not hard to check. The embedding function can be defined in the obvious way. That is, any point of S is mapped to itself. Any line ℓ of S is mapped to the line that contains ℓ and perhaps some new points. □

In general, the projective plane obtained in lemma 3.3.1 is infinite, but this is not always the case. The reader is left to devise an example which results in a finite plane.

3.4 Subplanes

A *subplane* Π' of a projective plane Π is a projective plane whose points are a subset of the points of Π and in which each line is a subset of a line of Π. We shall see later in the chapter how to obtain subplanes of planes, but for now we shall just examine some of their properties.

Theorem 3.4.1 (Bruck, 1963). *If Π' is a subplane of a finite projective plane Π and if the orders of Π and Π' are n and m respectively, then $\Pi \neq \Pi'$ implies*

$$n = m^2 \quad \text{or} \quad m^2 + m \leq n.$$

Proof. Suppose $\Pi \neq \Pi'$. Then clearly $m < n$. Let ℓ be a line of Π'. Then it has a point $p \notin \Pi'$. If there is a second line ℓ' on p meeting Π' in two or more points, then ℓ and ℓ' meet in Π' since Π' is projective, and since $p = \ell \cap \ell'$ we get $p \in \Pi'$ which is false. So all lines on p, except ℓ itself, meet Π' in at most one point.

However, each point of Π' is joined by a line to p, and there are $(m^2+m+1)-(m+1) = m^2$ such points not on ℓ. Therefore $n+1 \geq m^2+1$ or $n \geq m^2$.

If $n+1 > m^2+1$, there is a line on p not meeting Π' at all. Then all of the m^2+m+1 lines of Π' meet this line in distinct points (recall from above that two lines of Π' have intersection in Π'). Therefore $n+1 \geq m^2+m+1$ or $n \geq m^2+m$. □

Corollary. *If Π' is a subplane of order m in a finite projective plane Π of order $n = m^2$, then each line of Π contains at least one point of Π'.*

Theorem 3.4.1 gives us some important restrictions on the existence of projective planes inside other projective planes. For instance, the projective plane of order 3 cannot contain the Fano plane, as $3 \neq 2^2$ and $3 \ngeq 2^2+2$. The projective plane of order 9 may only contain projective planes of orders 2 and 3. (Note that we have not shown that the projective plane of order 9 *must* contain projective planes of these orders.)

A *Baer subplane* of a projective plane Π is a subplane Π' of Π such that every point of Π lies on a line of Π' and every line of Π meets Π' in at least one point.

In view of the above corollary and exercise 9 of section 3.11, if $n = m^2$, then Π' is a Baer subplane:

Lemma 3.4.2. *If Π' is a subplane of Π of order m and if Π has order $n = m^2$, then Π' is a Baer subplane of Π.*

Lemma 3.4.3. *If Π' is a Baer subplane of Π of order m and if Π has order n then either $n = m$ or $n = m^2$.*

The proof is left to the reader.

A *quadrangle* is a set of four points no three of which are collinear.

Axiom PP2 tells us that quadrangles always exist in projective planes.

Let a, b, c, d be the points of a quadrangle in a projective plane Π. We use a modification of the embedding process of the previous section to generate a projective subplane of Π from the quadrangle. Add the lines ab, ac, ad, bc, bd, cd; then the points of intersection in Π of these lines; then the lines determined in Π by this larger set of points; then the points of intersection in Π of this larger set of lines; and so on.

When we are finished this process, we have a projective plane which may in fact be Π. But if a proper projective subplane of Π exists we can find it if we are fortunate enough initially to choose a quadrangle inside it. Hence one way (a long and tedious one) of finding subplanes in a projective plane is by completing quadrangles as above.

3.5 Collineations in projective planes

Recall (see section 1.7) that a *collineation* is a 1–1 linear function mapping a space onto itself and such that f^{-1} is linear. We proved in lemma 1.7.4 that the set of all collineations of any near-linear space forms a group under composition of functions. We also saw (lemma 2.6.2) that, if the domain of f is a linear space and f is a 1–1 onto linear function, then f^{-1} is automatically linear.

In this section, we talk about collineations of projective planes. Consider the Fano plane Π of figure 1.1.1. The map f_1 defined by $f_1(4) = 4, f_1(3) = 3, f_1(6) = 6, f_1(0) = 2, f_1(5) = 1, f_1(2) = 0, f_1(1) = 5$ is a collineation of Π, as is the map f_2 defined by $f_2(1) = 4, f_2(2) = 0, f_2(4) = 5, f_2(0) = 6, f_2(5) = 1, f_2(6) = 2, f_2(3) = 3$.

A *central collineation* (also called a *perspectivity*) is a collineation f such that there is a point c, called the *centre*, with the property that $f(c) = c$, and $f(x) \in xc$ for every other point x.

In other words, a central collineation maps any point $x \neq c$ back onto the line on that point and the centre. The reader can check that f_2 above is not a central collineation, while f_1 is and has centre 6. The function defined below is also a central collineation of the Fano plane.

Example 3.5.1. $f(1)=1$, $f(3)=3$, $f(0)=0$, $f(2)=4$, $f(4)=2$, $f(5)=6$, $f(6)=5$. The reader can check that a centre is 1. In fact, 3 and 0 are also fixed by f but they are not centres.

Example 3.5.2. Let Π be the extended real plane. Each real point (x, y) will be mapped by f to $(2x, 2y)$. The new points have not been assigned co-ordinates. We will define $f(z)=z$ for each point at infinity z. Clearly, any line through $(0,0)$ gets mapped to itself while any other line with real points is moved. The line at infinity is fixed, however. (Note that any line is mapped either to itself or to a line parallel to itself.) The map f above is called a *dilation* or a *dilatation*.

The obvious question to ask is, is it possible for a collineation to have more than one centre? We answer this in the next lemma.

Lemma 3.5.1. *Unless the collineation f is the identity function, it cannot have more than one centre.*

Proof. Suppose that f has two centres c_1 and c_2. Let x be any point not on the line c_1c_2. Then $f(xc_1)=xc_1$ and $f(xc_2)=xc_2$. So $f(x) \in xc_1$ and $f(x) \in xc_2$. However, $xc_1 \cap xc_2 = x$ so that $f(x)=x$.

Let y be any point different from x. The line xy has at least three points (lemma 3.2.1) and since only one of these is on c_1c_2, at least two points of the line xy are fixed by f and so $f(xy)=xy$. So $f(y) \in xy$ and x is a centre.

If we now repeat the first argument using c_1 and x as centres, we can show that all the points on c_1c_2 are also fixed and so f is the identity function. □

Recalling the notion of duality in projective planes it makes sense to introduce the next definition.

An *axial collineation* is a collineation which fixes every point of a particular line. This line is called an *axis*.

Each of the central collineations above is an example of an axial collineation. In fact, we prove in the next lemma that a central collineation always has an axis.

Lemma 3.5.2. *A non-identity central collineation f has one and only one axis.*

Proof. Let ℓ and m be two axes of f. Any line not on $\ell \cap m$ meets ℓ and m in different points and hence is mapped to itself. Any point x not on ℓ or m is therefore on at least two distinct lines, each of which maps to itself. So x is fixed, and f is the identity, a contradiction.

Any central collineation has lines which map to themselves: for instance, each line on the centre c is mapped to itself. If there is a line ℓ not on c which maps to itself, then each point x of ℓ has the property that $f(x) \in \ell \cap xc = x$; that is, $f(x) = x$. Hence ℓ is an axis.

Finally, suppose that all fixed lines are on c. Let ℓ be any line not on c, and let $a = \ell \cap f(\ell)$. Since $f(a) \in ac \cap f(\ell)$, a is fixed. Let b be any point of $ac \setminus \{a, c\}$. Suppose $f(b) \neq b$. Let h be any line on b, $h \neq ac$, and let $x = h \cap f(h)$. Thus, $x \notin ac$ and since $x \in xc \cap f(h)$, x is fixed. Hence the line ax is fixed by f and $c \notin ax$ which is a contradiction. So $b = f(b)$ and the line ac is an axis. □

A central collineation with axis on the centre is called an *elation*.

A central collineation with axis not on the centre is called a *homology*.

Giving dual arguments in the proof of lemma 3.5.2 shows that a non-identity axial collineation has a unique centre. (See exercise 18 of section 3.11.)

If f has centre c and axis ℓ, we call it a *(c, ℓ)-collineation.*

We urge the reader to check that the set of all (c, ℓ)-collineations is a group. (See exercise 17 of section 3.11.) Let us now consider the question of the number of (c, ℓ)-collineations. Once a centre and axis are given, is the collineation uniquely determined? The answer to this is no. Consider example 3.5.2 above, where Π is the extended real plane. Any map $f:(x, y) \to (kx, ky)$, k a non-zero real and where f maps each point at infinity to itself, is a $((0,0), \ell_\infty)$-collineation. The next result tells us when a central collineation is completely determined.

Lemma 3.5.3. *A (c, ℓ)-collineation is uniquely determined by its centre c, its axis ℓ and its effect on one other point $x \neq c$, $x \notin \ell$.*

Proof. Clearly, $f(x) \notin \ell$.

Suppose $y \notin xc, \ell$. Let $xy \cap \ell = z$, a fixed point. So $f(y) = cy \cap f(x)z$. (See figure 3.5.1.)

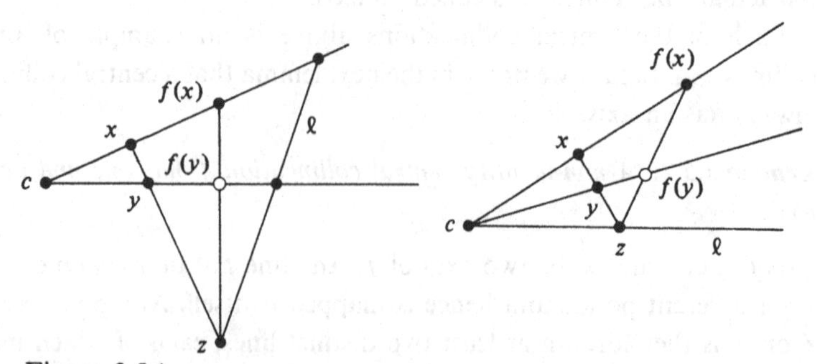

Figure 3.5.1.

So the effect of f on all points not on cx is known. For y on cx, choose $x' \notin cx$, ℓ, and use x' and $f(x')$ in the above discussion instead of x and $f(x)$, to determine $f(y)$. □

Corollary. *A non-identity (c,ℓ)-collineation fixes no points other than c and points of ℓ.*

Proof. Consider, in the above proof, the case $f(x)=x$. This forces $f(y)=y$ for all $y \notin xc$, and hence for all y. □

A plane is *(c,ℓ)-transitive* for some fixed point c and fixed line ℓ if, for any pair of distinct points x and x' such that neither point is in $\{c\} \cup \ell$ and c, x and x' are collinear, there is a (c,ℓ)-collineation mapping x to x'.

It is easy to see that the Fano plane is (c,ℓ)-transitive for any c and ℓ.

Consider the extended real plane Π, and choose $c=(0,0)$ and $\ell=\ell_\infty$. We shall show that Π is (c,ℓ)-transitive. Let (x,y) and (x',y') be distinct points with c, (x,y) and (x',y') collinear, (x,y), $(x',y') \notin \ell$, $(x,y), (x',y') \neq c$. Checking slopes on the line segments $(0,0)$ to (x,y) and $(0,0)$ to (x',y') we have $yx'=xy'$, supposing $x \neq 0$. We must find a (c,ℓ)-collineation mapping (x,y) to (x',y'). We try $f:(x,y) \to (rx,ry)$ which is a generalization of the function used in example 3.5.2. We must determine $r \neq 0$. Clearly f fixes c and each point of ℓ_∞. Since $rx=x'$ and $ry=y'$, we should choose $r=x'/x$ and $r=y'/y$ so that we need $x'/x=y'/y$; or $x'y=y'x$ which is the case as long as $x \neq 0$, and $y \neq 0$. If $x=0$, then $x'=0$ while y and y' are not zero. In this case we can choose $r=y'/y$. If $y=0$, then $y'=0$ while neither x nor x' is zero, so that we can choose $r=x'/x$.

Lemma 3.5.4. *If Π is (c,ℓ)-transitive and f is any collineation of Π, then Π is also $(f(c), f(\ell))$-transitive.*

Proof. Let $f(x)$ and $f(x')$ be points not in $\{f(c)\} \cup f(\ell)$ such that $f(x)$, $f(x')$ and $f(c)$ are collinear. We must find an $(f(c), f(\ell))$-collineation mapping $f(x)$ to $f(x')$.

As f is a collineation, x, x' and c are collinear. Since Π is (c,ℓ)-transitive there is a (c,ℓ)-collineation g mapping x to x'. Let $h=f \circ g \circ f^{-1}$ which is a collineation fixing $f(c)$ and each point of $f(\ell)$. It also maps $f(x)$ to $f(x')$. It follows from this last fact that h maps lines on $f(c)$ to themselves. So h is the collineation we are looking for. □

Π is said to have a *complete set of central collineations (perspectivities)* if it is (c,ℓ)-transitive for any c and ℓ.

3.6 The Desargues configuration

Let f be a (c,ℓ)-collineation. Choose x, y and z not in $\{c\} \cup \ell$ and such that c, x, y and z form a quadrangle. Let x', y' and z' be $f(x)$, $f(y)$ and

$f(z)$ respectively. Let $u = xy \cap x'y'$, $v = xz \cap x'z'$, $w = yz \cap y'z'$. (See figure 3.6.1.) Clearly $u \neq c$. But $u \in xy$ implies $f(u) \in f(xy) = x'y'$ and also $f(u) \in uc$. So u is a fixed point. Similarly, v and w are fixed points and so all three points are on ℓ. Figure 3.6.1 shows the homology case. We leave the figure in the elation case as an exercise. (See exercise 23 of section 3.11.)

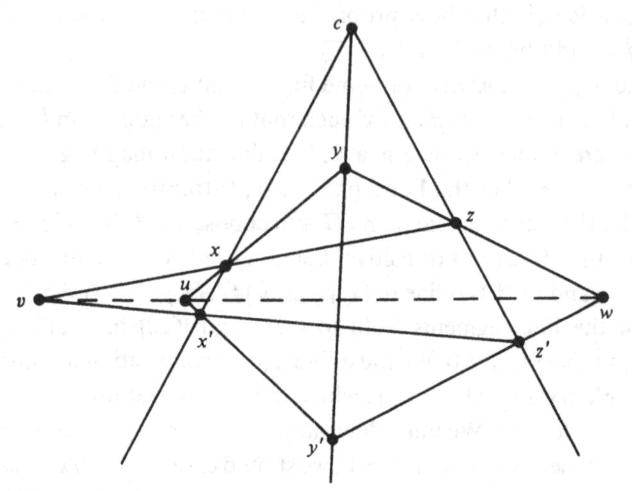

Figure 3.6.1.

Any system of ten points as shown in figure 3.6.1, whether they come from a (c, ℓ)-collineation or not, is called a *Desargues* (or *Desarguesian*) configuration. We emphasize that c, x, y, z, x', y', z' are distinct points such that c, x, x' and $c, y\, y'$ and c, z, z' are collinear, and also such that x, y, z and x', y', z' are 'triangles'. That is, x, y and z (x', y' and z') are not collinear. And finally, the points $xy \cap x'y'$, $xz \cap x'z'$ and $yz \cap y'z'$ are collinear.

Lemma 3.6.1. *Except for the Fano plane, a projective plane Π with a non-identity (c, ℓ)-collineation has a Desargues configuration induced by the collineation.*

The proof was given at the beginning of this section. Note that in case Π is the Fano plane, a non-identity (c, ℓ)-collineation must be an elation and it is impossible to choose x, x', y, y', z, z' as required.

Suppose that for any appropriate choice of c, x, y, z, x', y' and z', the points u, v and w as above are always collinear. Then Π is said to be *Desarguesian.* We can now say that the Fano plane is Desarguesian – in a vacuous sense. We cannot choose c, x, y, z, x', y' and z' as in figure 3.6.1 and so it is impossible for us to choose them in such a way that u, v and w are not collinear.

If Π is Desarguesian, this is in fact equivalent to having a complete set of central collineations. (See Hughes and Piper (1973) for a proof, for example.) If it is not Desarguesian, we call it *non-Desarguesian*. There are several possible categories of non-Desarguesian planes.

(*a*) There may be some set of ten points satisfying the above, but no (c,ℓ)-collineations at all. (*All* finite projective planes contain Desargues configurations: Hughes and Piper (1973).)

(*b*) There may be some set of ten points satisfying the above and at least one (c,ℓ)-collineation, where ℓ is the line on u, v and w.

(*c*) For a particular choice of c and ℓ and for all choices of x, x', y and z as above, the other points of the configuration can always be found. (Note that this means that Π is (c,ℓ)-transitive.)

(*d*) If (c) is satisfied for each $c\in\ell$, we call Π a *translation plane with respect to* ℓ.

We stated in the above paragraph that Π Desarguesian is equivalent to Π having a complete set of central collineations. We shall prove this here in one direction as we shall need the result later in the chapter.

Lemma 3.6.2. *If Π has a complete set of central collineations then Π is Desarguesian.*

Proof. Let c, x, x' y, y' z, z' be chosen appropriately (see figure 3.6.1). As Π has a complete set of central collineations, there is a (c, uv)-collineation f, where u, v (and w) are defined as in figure 3.6.1, which maps x to x'. Hence $f(xu) = x'u$ so that $f(y)\in f(x'u)$. But $f(y)\in cy$ which forces $f(y) = y'$. Similarly, $f(z) = z'$. Now $f(w)\in cw\cap f(yz) = cw\cap y'z' = \{w\}$ so that w is a fixed point. Either $w\in uv$, the axis, or $w = c$, by the corollary to lemma 3.5.3. In the former case, we have completed the proof. If $w = c$, then $c\in xx'\cap xz$ forces $c = x$, a contradiction. □

3.7 Construction of projective planes from vector spaces

In this section we show how any field can be used to construct a projective plane.

A *field* is set F with two binary operations, usually denoted by $+$ and $\cdot$ (plus and times) such that F is a commutative group (see section 1.7) under $+$ with identity usually denoted by 0, $F\backslash\{0\}$ is a commutative group under $\cdot$ with identity usually denoted by 1, and also, $a\cdot(b+c) = a\cdot b + a\cdot c$ and $(a+b)\cdot c = a\cdot c + b\cdot c$. (See exercise 25 of section 3.11.) This system is called a *skew-field* if the commutative condition under times: $a\cdot b = b\cdot a$ is not required. We do not need this condition at all in this section, but it simplifies things if we do not have to worry about it. In exercise 37 of section 3.11 we ask the reader to check that we do not need it.

We usually write *ab* instead of $a \cdot b$.

The real numbers and complex numbers are fields, as can be checked by the reader. There are many finite fields. For example the set $\{0, 1, 2\}$ with addition $\oplus$ and multiplication $\odot$ taken modulo 3 forms a field. That is, $a \oplus b$ or $a \odot b$ is the element of $\{0, 1, 2\}$ which is the remainder of ordinary integer addition $a + b$, or multiplication $a \cdot b$ on division by 3. So $2 \oplus 1 = 0, 2 \odot 2 = 1$ and so on. We have seen this modular arithmetic before – in connection with the Fano plane in section 1.1.

Not every number k produces a field in this way. If k is a prime number, we do get a field. If k is a prime power, say $k = p^n$, p a prime and n a positive integer, we can also construct a field with k elements as follows. Let $f(x)$ denote a polynomial of degree n with coefficients taken from a field F with a prime number p of elements and operations defined modulo p. Suppose this polynomial is irreducible in F (that is, cannot be factored non-trivially in F). However, $f(x)$ has roots in some larger field. (For the details see Hirschfeld (1979).) Let one such root be a. Define $F(a) = \{\alpha_{n-1}a^{n-1} + \alpha_{n-2}a^{n-2} + \cdots + \alpha_0 | \alpha_i \in F\}$. Since $f(a) = 0$, powers of a higher than $n - 1$ can be rewritten in terms of powers of a lower than or equal to $n - 1$. It is possible to check that $F(a)$ is a field where addition and multiplication are defined as usual for polynomials, treating a as a 'variable'. Moreover, it is easy to see that this field has p^n elements.

The *characteristic* of a finite field F is the smallest positive integer p such that $px = 0$ for all x in F.

It is possible to show that every finite field has a characteristic p, that p is always a prime, and also that any field of characteristic p contains the field of p elements in which the operations are modulo p.

We use the notation GF(k)† for a finite field with k elements.

To construct a projective plane, our general method of attack is to take a three-dimensional vector space V over a field F where each point is a triple (a, b, c) of field elements a, b and c. If the field has k elements, there are k^3 such points, each line has k points and each plane through the origin is on $k + 1$ lines through the origin. (The reader should verify these statements.)

The lines through the origin are the one-dimensional subspaces. That is, they are the sets $\{\alpha v | \alpha \in F\}$ where v is in V. The planes through the origin are the two-dimensional subspaces: $\{\alpha v + \beta w | \alpha, \beta \in F\}$ where v and w are fixed elements of V and $w \neq \alpha v$ for any $\alpha \in F$. In other words, w is not on the line determined by v.

† The letters GF stand for Galois field, as the whole theory of finite fields is due to the French mathematician Galois (Bell, 1967).

As an example, let us consider the three-dimensional vector space V over GF(2). There are $2^3 = 8$ points, each having the form (a, b, c) where a, b and c are either 0 or 1. This vector space is represented by figure 3.7.1. The planes on $(0,0,0)$ are

$(0,0,0)$, $(1,0,0)$, $(0,1,0)$, $(1,1,0)$;
$(0,0,0)$, $(1,0,0)$, $(0,0,1)$, $(1,0,1)$;
$(0,0,0)$, $(1,0,0)$, $(0,1,1)$, $(1,1,1)$;
$(0,0,0)$, $(0,1,0)$, $(1,1,1)$, $(1,0,1)$;
$(0,0,0)$, $(0,1,0)$, $(0,1,1)$, $(0,0,1)$;
$(0,0,0)$, $(0,0,1)$, $(1,1,1)$, $(1,1,0)$;
$(0,0,0)$, $(1,1,0)$, $(0,1,1)$, $(1,0,1)$.

In general, a plane has k^2 points, so a formula for the number of planes through the origin is given by

$$\frac{(k^3-1)(k^3-k)}{(k^2-1)(k^2-k)} = k^2 + k + 1.$$

We leave the details of this as an exercise.

It is also true that any two planes through the origin meet in a line. We can observe this quickly by examining the above planes for GF(2). We shall need this fact in the proof of lemma 3.7.1, along with the fact that two intersecting lines of a vector space determine a unique plane. For more information on vector spaces, the reader should see Curtis (1967).

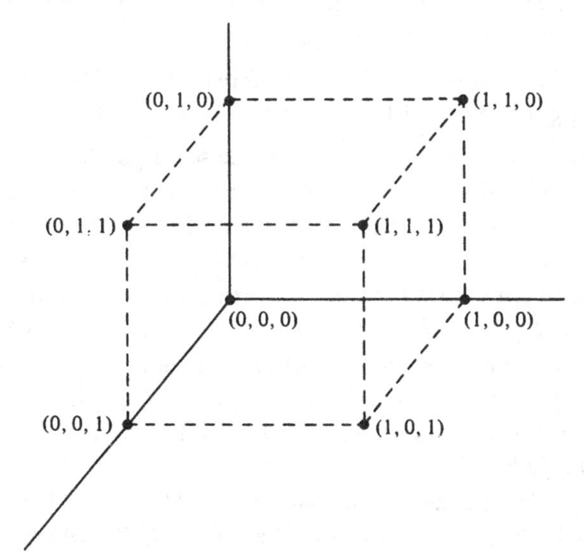

Figure 3.7.1.

Subspaces are closed under addition, and so the addition of two points of a line (plane) on $(0,0,0)$ is another point on the line (plane). For instance in GF(5), $(1,1,2)$ and $(2,2,4)=2(1,1,2)$ are collinear with $(0,0,0)$, so $(1,1,2)+(2,2,4)=(3,3,1)=3\cdot(1,1,2)$ is also on the line. The other point is $(4,4,3)=4\cdot(1,1,2)$.

To produce the projective plane, each line of V through the origin becomes a point, and each plane of V through the origin becomes a line. A line that is on a plane in V becomes a point of the corresponding line in the new system. We prove in the next lemma that this is indeed a projective plane.

Notation. We let $\Pi(V)$ denote the system constructed as above from V.

Lemma 3.7.1. *The structure $\Pi(V)$ is a projective plane.*

Proof. Any two points in $\Pi(V)$ are on a unique line as the corresponding lines in V are on a unique plane. Each line of $\Pi(V)$ has at least two points as each plane of V through the origin is on at least two lines through the origin. Any two lines of $\Pi(V)$ meet as the corresponding planes of V, already meeting in the origin, must meet in a line. Finally we can always find four lines of V through the origin such that no three are coplanar; for example the lines on $\{\{0,0,0\},\{1,0,0\}\}$, $\{\{0,0,0\},\{0,1,0\}\}$, $\{\{0,0,0\},\{0,0,1\}\}$, and $\{\{0,0,0\},\{1,1,1\}\}$ are always available using any field. These four lines correspond to four points of $\Pi(V)$ no three collinear. Hence $\Pi(V)$ is a projective plane. □

Lemma 3.7.2. *The projective plane $\Pi(V)$ produced as above from a vector space V over a finite field* GF(q) *has order q.*

Proof. This follows from the statement above, that each plane of V through the origin is on $q+1$ lines through the origin. □

Each point of $\Pi(V)$ can be considered as the set of points of the line it corresponds to, or as the set of multiples of a particular point of that line, or even as just a particular point of that line. Hence, this produces a co-ordinatization of the projective plane. Each point may be assigned a triple of field elements. To distinguish between a vector space point and a projective plane point, we shall use $[a,b,c]$ for the latter. So $[a,b,c]$ and $[ak,bk,ck]$ are the same point of $\Pi(V)$ for any non-zero field element k. Notice that the point $[0,0,0]$ does not exist in $\Pi(V)$, so that any point $[a,b,c]$ with $a\neq 0$ can be identified with $[1,b',c']$, and any other point with $[0,1,c']$ or $[0,b',1]$.

As an example, let us construct the Fano plane Π from a field. As Π has order 2 we need a field of two elements. We can use GF(2) which has the elements 0 and 1 and addition and multiplication are modulo two.

Example 3.7.1. Consider the vector space V over GF(2) where the points are $(0,0,0)$, $(1,0,0)$, $(0,1,0)$, $(0,0,1)$, $(1,1,0)$, $(1,0,1)$, $(0,1,1)$ and $(1,1,1)$. All lines have precisely two points, so there are seven lines through the origin. Each plane through the origin contains four points. The projective plane $\Pi(V)$ therefore has seven points and seven lines. Each line has three points and each point is on three lines. By lemma 3.2.2 this is the Fano plane. We redraw it below in figure 3.7.2 using the above field labelling. Notice that in this case, as lines of V have only two points, it is possible to identify the non-zero points of V with the points of $\Pi(V)$.

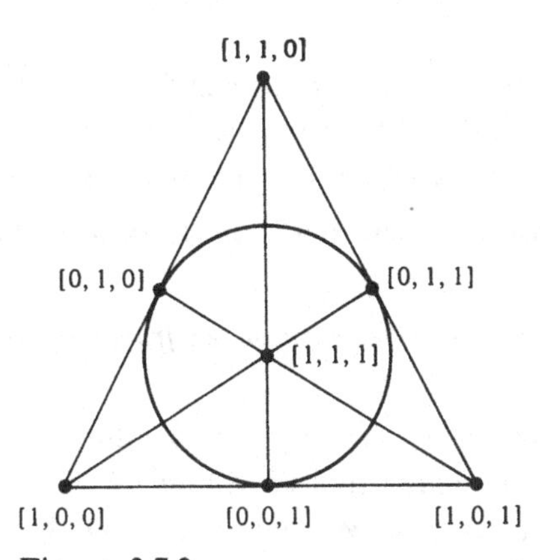

Figure 3.7.2.

We note also that the general construction above tells us that projective planes of order k exist for any k a prime power.

As subspaces in a vector space V over the field F can be described in terms of equations, the same should be true in $\Pi(V)$ and therefore in planes through the origin in V.

Any plane of V passing through the origin has the form $ax+by+cz=0$ for a, b and c fixed field elements. The point (x,y,z) is on this plane if and only if it satisfies the equation. Let us introduce the notation $\langle a,b,c\rangle$ for the line of $\Pi(V)$ corresponding to the above plane of V. Notice that $\langle a,b,c\rangle = \langle a',b',c'\rangle$ precisely when, for some non-zero field element $r, a=ra'$, $b=rb'$, $c=rc'$. In $\Pi(V)$, we can now state that the point $[x,y,z]$ is on the line $\langle a,b,c\rangle$ if and only if $ax+by+cz=0$. We can think of the latter equation as a vector product $\langle a,b,c\rangle \begin{bmatrix} x \\ y \\ z \end{bmatrix} = 0.$

Recalling that the subspace defined by two lines through the origin consists of sums of multiples of the points on these lines, we see that, in $\Pi(V)$, the line on two points $[x,y,z]$ and $[x',y',z']$ is the set of points of the form $[x, y, z]\alpha + [x', y', z']\beta = [x\alpha + x'\beta, y\alpha + y'\beta, z\alpha + z'\beta]$ for any field elements α and β not both zero.

Since the choice of axes in V is arbitrary, we are allowed to make some assumptions about co-ordinates when working in $\Pi(V)$. For instance, a general quadrangle in $\Pi(V)$ may be assumed to have the co-ordinates $[1,0,0]$, $[0,1,0]$, $[0,0,1]$ and $[1,1,1]$.

We shall find it useful later to have the following result handy.

Lemma 3.7.3. *Three points in $\Pi(V)$ are collinear if and only if the matrix whose elements are the co-ordinates of the points is singular (has zero determinant).*

Proof. Let the points under consideration be x, y and z. Then z lies on the line xy if and only if $z = x\alpha + y\beta$ for some field elements α and β. Hence the corresponding matrix of columns of x, y and z is singular. □[†]

We conclude this section with a result that is important enough to be labelled a theorem.

Theorem 3.7.4. *Any projective plane $\Pi(V)$ constructed from a vector space V over a field F is Desarguesian.*

Proof. Choose distinct points c, x, y, z and x', y', z' such that x, y, z and x', y', z' form triangles and x, x', c are collinear, y, y', c are collinear and z, z', c are collinear. (See figure 3.7.3.) We must show that the points $xy \cap x'y' = u$, $xz \cap x'z' = v$ and $yz \cap y'z' = w$ are collinear.

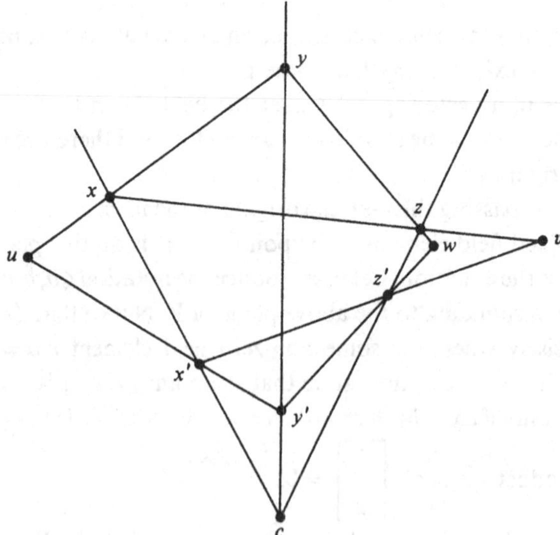

Figure 3.7.3.

[†] The reader is warned that, if F is not commutative, determinants do not make sense.

As was mentioned above, we may arbitrarily allot c, x, y and z the coordinates $[1,0,0]$, $[0,1,0]$, $[0,0,1]$ and $[1,1,1]$ respectively. Then the lines xc, yc and zc are respectively $\langle 0,0,-1\rangle$, $\langle 0,1,0\rangle$ and $\langle 0,1,-1\rangle$. (This can be shown directly by letting the line be $\langle a,b,c\rangle$ and getting solutions for a, b and c by putting x and c, or y and c, or z and c into the equation of the line. Alternatively, see exercise 34 of section 3.11.)

Since $x'\in xc$, $x'=[1,0,0]\alpha+[0,1,0]\beta=[\alpha,\beta,0]$ for some non-zero field elements α and β. We can therefore set $x'=[1,t,0]$ for $t\neq 0$. Similarly, $y'=[1,0,s]$, $s\neq 0$, and $z'=[r,1,1]$, $r\neq 1$.

The fact that $u\in xy$ implies $u=[0,1,a]$ for $a\neq 0$. And the fact that $u\in x'y'$ implies $u=[0,1,-st^{-1}]$. So $a=-st^{-1}$. Similar arguments show that $v=[1,b,1]=[1,t(1-r)+1,1]$, $b\neq 0$, and so $b=t(1-r)+1$; and $w=[1,1,c]=[1,1,s(1-r)+1]$, $c\neq 0$, and so $c=s(1-r)+1$.

It remains only to show that u, v and w are collinear, and we do this by using lemma 3.7.3. But

$$\det\begin{pmatrix} 0 & 1 & 1 \\ 1 & t(1-r)+1 & 1 \\ -st^{-1} & 1 & s(1-r)+1 \end{pmatrix} =$$
$$-(s(1-r)+1+st^{-1})+(1+st^{-1}t(1-r)+st^{-1})=0. \quad \square$$

3.8 The Pappus configuration

As was mentioned in section 3.7, if we drop the condition that a field be commutative under $\cdot$ (but keep all of the other conditions), the new system is known as a *skew-field*. The reader was asked to note that the construction of $\Pi(V)$ in section 3.7 is valid even if F is a skew-field.

Suppose that $\Pi(V)$ is a projective plane constructed over a skew-field F. Then it is possible to tell from the geometry of $\Pi(V)$ precisely when F is commutative under $\cdot$ and therefore a field. This is known as Pappus' theorem. Note that throughout we shall use scalar multiplication on the right hand side.

Theorem 3.8.1. *Let Π be a projective plane co-ordinatized over a skew-field F. Let x, y, z and x', y', z' be sets of three distinct collinear points on distinct lines, such that no one of these points is the intersection of these two lines. Let $u=xy'\cap x'y$, $v=xz'\cap x'z$, $w=yz'\cap y'z$. Then u, v and w are collinear for all choices of x, y, z, x', y', z' as above, if and only if F is a field.*

Proof. Assume that $x, y, z, x', y', z', u, v$ and w are as given. (See figure 3.8.1.)

As x, y, x' and y' form a quadrangle, we may label them $[1,0,0]$, $[0,1,0]$, $[0,0,1]$ and $[1,1,1]$ respectively. Then $z\in xy$ so it has the form $[1,s,0]$, $s\neq 0$. Similarly, $z'\in x'y'$ implies $z'=[1,1,1+t]$, $t\neq 0$.

Since $u \in xy' \cap x'y$, $u = [\alpha + 1, 1, 1] = [0, 1, \beta]$ for skew-field elements α and β. Hence we may set $u = [0, 1, 1]$.

From $v \in xz' \cap x'z$ we obtain $v = [1, s, \alpha] = [\beta, 1, 1 + t]$ for some skew-field elements α and β. So we may set $v = [1, s, (1 + t)s]$.

Finally $w \in yz' \cap y'z$ implies $w = [1, \alpha, 1 + t] = [1 + \beta, 1 + s\beta, 1]$ for some skew-field elements α and β. It follows that $\alpha = (1 + s\beta)(1 + t) = 1 + t + s\beta + s\beta t$ while $1 = (1 + \beta)(1 + t) = 1 + \beta + t + \beta t$ implies $-t = \beta + \beta t$ so $-st = s\beta + s\beta t$. Hence $\alpha = 1 + t - st$. So we may write $w = [1, 1 + t - st, 1 + t]$.

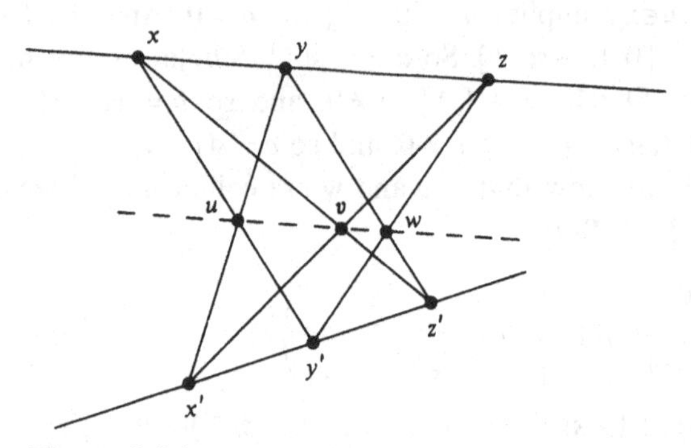

Figure 3.8.1.

The line on points u and v has co-ordinates $\langle ts, 1, -1 \rangle$. The point w is on this line if and only if

$$\langle ts, 1, -1 \rangle \begin{bmatrix} 1 \\ 1 + t - st \\ 1 + t \end{bmatrix} = 0;$$

that is, if and only if $ts + (1 + t - st) - (1 + t) = 0$. This is true precisely when $ts = st$.

Since s and t vary over all non-zero skew-field elements as z and z' vary over the points of the lines xy and $x'y'$, we see that the commutativity of F is a necessary and sufficient condition for u, v and w to be collinear. □

The configuration of nine points and nine lines described in the theorem is called a *Pappus configuration.* And a projective plane in which u, v and w constructed as above are always collinear is said to be *Pappian.*

It turns out that one can prove that if a projective plane is Pappian then it is Desarguesian. A number of proofs of this exist (see, for example, Hartshorne (1967) or Szmielew (1983)). The one we give is due to Cronheim (1953).

Theorem 3.8.2. *Let Π be a Pappian projective plane. Then Π is Desarguesian.*

Proof. In the following argument it is very convenient to use the idea of permutation of subscripts. We therefore use a notation which is slightly different from that we have used to this point.

Let $c, x_1, y_1, x_2, y_2, x_3, y_3$ be distinct points such that the triples c, x_1, y_1 and c, x_2, y_2 and c, x_3, y_3 are collinear and the lines through them are distinct.

We note first of all the following property. Suppose that there is no permutation (i, j, k) of the numbers $1, 2, 3$ such that simultaneously x_i, y_j, y_k and x_i, x_j, y_k are non-collinear. Then we claim that either x_a, y_b, y_c are collinear for all permutations (a, b, c) or y_a, x_b, x_c are collinear for all permutations (a, b, c). To see this, we may assume without loss of generality that y_1, x_2, x_3 are non-collinear. It follows then that x_2, y_1, y_3 are collinear and x_3, y_1, y_2 are collinear. Clearly, then x_2, x_1, y_3 cannot be collinear. Thus, by hypothesis, x_1, y_2, y_3 are collinear. That is, x_a, y_b, y_c are collinear for all permutations (a, b, c).

Now let $u_i = x_j x_k \cap y_j y_k$ for each $i = 1, 2, 3$, and (i, j, k) a permutation of $1, 2, 3$. We need to show that u_1, u_2, u_3 are collinear. We distinguish two cases.

Case 1. Suppose there is a permutation (i, j, k) of 1, 2, 3 such that x_i, y_j, y_k are non-collinear and x_i, x_j, y_k are non-collinear. We may assume for instance that x_1, y_2, y_3 are non-collinear and x_1, x_2, y_3 are non-collinear. Define $v = x_1 x_2 \cap y_2 y_3$. It is easy to see that v is distinct from each x_i and from each y_i. Moreover $cx_1 \neq cx_2$ implies $v \neq c$. Consider the triples of collinear points c, x_3, y_3 and v, x_1, x_2 which form a Pappus configuration. Thus $u_1 = y_3 v \cap x_2 x_3$, $x = cx_2 \cap x_1 y_3$ and $z = cv \cap x_1 x_3$ are collinear. Similarly, the triples of collinear points c, x_1, y_1 and v, y_2, y_3 form a Pappus configuration leading to u_3, x and $z' = y_1 y_3 \cap cv$ collinear. Now v, z, z' are collinear, as are x, x_1, y_3, and all points are distinct. Moreover, it is easy to see that none of v, z, z' lies on $x_1 y_3$. We thus have a Pappus configuration and so $u_1 = xz \cap vy_3$, $u_2 = y_3 z' \cap x_1 z$, $u_3 = x_1 v \cap z'x$ are collinear as desired.

Case 2. Suppose there is no permutation (i, j, k) of $1, 2, 3$ such that x_i, y_j, y_k are non-collinear and x_i, x_j, y_k are non-collinear. It follows from the remarks preceding case 1 that either x_a, y_b, y_c are collinear for all permutations (a, b, c) or y_a, x_b, x_c are collinear for all permutations (a, b, c). We assume without loss of generality that x_1, y_2, y_3 are collinear, x_2, y_1, y_3 are collinear and x_3, y_1, y_2 are collinear.

Note that no u_i equals any x_j or y_k. Define $v_1 = cx_1 \cap y_3 u_3$ and $v_2 = cx_2 \cap y_3 u_3$. Note that $y_3 u_3 \neq y_1 y_3$, $y_2 y_3$, $x_3 y_3$. The points u_3, $v_1, v_2, x_1, x_2, x_3, y_1, y_2, y_3, c$ are therefore all distinct. Also, the two collinear triples c, x_3, y_3 and x_1, x_2, u_3 lie on different lines. The Pappus property now implies that $u_2 = x_1x_3 \cap x_2y_3$, $v_1 = u_3y_3 \cap cx_1$ and $y_2 = cx_2 \cap x_3u_3$ are collinear. The same triples also imply, using Pappus, that $u_1 = x_2x_3 \cap x_1y_3$, $v_2 = cx_2 \cap u_3y_3$ and $y_1 = cx_1 \cap x_3u_3$ are collinear. Finally, consider the collinear triples x_1, y_1, v_1 and x_2, y_2, v_2. Since $u_1 = x_1y_2 \cap y_1v_2$, $u_2 = y_1x_2 \cap v_1y_2$ and $u_3 = v_1v_2 \cap x_1x_2$, the Pappus property implies that these three points are collinear as desired. □

3.9 Projective spaces

So far we have restricted ourselves to projective planes, which are linear spaces of dimension 2 with some additional properties. We should like to introduce higher-dimensional spaces which are somehow generalizations of projective planes. We make the following definition.

A *projective space* S is a linear space with the property that any two-dimensional subspace is a projective plane.

In view of exercise 34 of section 2.7, we have the following.

Lemma 3.9.1. *Any subspace of a projective space is a projective space.*

Notice that $\varnothing$, a point and a line are trivial examples of a projective space. We shall usually restrict our attention to non-trivial examples.

Lemma 3.9.2. *Any two lines of a projective space have the same number of points.*

Proof. Let ℓ and ℓ' be distinct lines. If they are in some plane, then the result is immediate. If not, let $p \in \ell$, $p' \in \ell'$. (Clearly, p cannot be p'.) Let $h = pp'$. Then ℓ and h are coplanar and so have the same number of points; while ℓ' and h are also coplanar and have the same number of points. □

Let the number of points per line be $k+1$, if the number is finite. Then we say S has *order* k. Clearly, k is the order of any projective plane in S.

Lemma 3.9.3. *Let V be any subspace of the projective space S and let $p \notin V$. Then the subspace $\langle V \cup \{p\} \rangle$ is the set of points on all lines pq for $q \in V$.*

Proof. Let X be the set of points contained in lines on p which meet V. We shall show that X is a subspace. Since it will clearly be the smallest subspace on V and p, the result will follow. We may assume $V \neq \varnothing$.

Let q and r be points of X. We must show $qr \subseteq X$. If p, q and r are collinear then, by definition of X, $qr = pq \subseteq X$. Suppose p, q and r are not collinear. Let $pq \cap V = q'$ and $pr \cap V = r'$. (Note that it may be the case that $q = q'$ or $r = r'$.) Now $\Pi = \langle \{p\} \cup \{q\} \cup \{r\} \rangle$ is a plane containing the distinct points q' and r'. So the line $q'r'$ is in Π, and also in V. Finally, let s be *any* point of qr. The line ps is in Π and, since Π is a projective plane, ps meets $q'r'$ in a point s'. Thus $s' \in V$ and so $s \in X$. □

Lemma 3.9.4. *A projective space S satisfies the exchange property.*

Proof. Suppose x and y are points of S and X is a subset of the points of S such that $x \notin \langle X \rangle$, $x \in \langle X \cup \{y\} \rangle$. Without loss of generality, we may suppose that $X = \langle X \rangle$ is a subspace.

By lemma 3.9.3, $x \in \langle X \cup \{y\} \rangle$ implies that x is on some line yp, $p \in X$, and clearly we may assume $x \neq p$. Thus y is on the line xp and so $y \in \langle X \cup \{x\} \rangle$. □

Corollary 1. *If the projective space S has finite dimension n, then any set of $n+1$ points of S not all in an $(n-1)$-dimensional subspace, form a basis for S.*

Proof. This follows from lemmas 3.9.4 and 2.4.5. □

Corollary 2. *The hyperplanes of an n-dimensional projective space are precisely the $(n-1)$-dimensional subspaces.*

Proof. This also follows from lemmas 3.9.4 and 2.4.5. □

Let H be a hyperplane of the projective space S. (See section 2.5.) By lemma 3.9.1, H is a projective space.

It is now possible to count the number of points in an n-dimensional projective space (projective n-space) of order k. Recall that a two-dimensional projective space (a projective plane) of order k has $k^2 + k + 1$ points. Consider a three-dimensional projective space S, that is a projective space generated by four points not all in a plane. Let the points be p, q, r and s. Letting $V = \langle \{p\} \cup \{q\} \cup \{r\} \rangle$, a plane, lemmas 3.9.3 and 3.9.4 tell us that $S = \langle V \cup \{s\} \rangle$ is the set of all points on lines st where $t \in V$. As there are $k^2 + k + 1$ points in V, this makes $(k^2 + k + 1)k + 1 = k^3 + k^2 + k + 1$ points altogether in S.

We can now prove inductively that an n-dimensional projective space S of order k has $k^n + k^{n-1} + \cdots + k + 1$ points. Letting $S = \langle \{p_1\} \cup \{p_2\} \cup \cdots \cup \{p_{n+1}\} \rangle$ and $H = \langle \{p_1\} \cup \{p_2\} \cup \cdots \cup \{p_n\} \rangle$ an $(n-1)$-dimensional subspace by lemma 3.9.4, we have by induction that the number of points of H is $k^{n-1} + k^{n-2} + \cdots + k + 1$. Then, by lemma 3.9.3, S has $(k^{n-1} + k^{n-2} + \cdots + k + 1)k + 1 = k^n + k^{n-1} + \cdots + k + 1$ points. We summarize this in the next lemma.

Lemma 3.9.5 *An n-dimensional projective space of order k has $k^n + k^{n-1} + \ldots + k + 1$ points, $n \geq 0$.*

Note that the cases $n = 0$ and $n = 1$ are trivial.

Lemma 3.9.6. *A proper subspace H of a projective space S is a hyperplane if and only if each line of S meets H in at least one point.*

Proof. Suppose every line meets H. Let $p \notin H$. Then for all $q \neq p$, the line pq meets H by assumption. Hence $S \subseteq \langle H \cup \{p\} \rangle$, or $S = \langle H \cup \{p\} \rangle$

for *any* $p \notin H$. If H were not a hyperplane, we could find a subspace V such that $H \subsetneq V \subsetneq S$ and then, choosing $P \in V \backslash H$, we have $\langle H \cup \{p\} \rangle \subseteq V \subsetneq S$ contradicting what we have just said. Therefore H is a hyperplane.

Assume H is a hyperplane. Lemma 3.9.3 implies that the set of points on lines of the form pq, $q \in H$ and p a fixed point not in H, is a subspace. As H is a hyperplane, $\langle H \cup \{p\} \rangle = S$ and so all lines meet H. □

Projective spaces, like projective planes, have many nice properties. To discover these properties, it often suffices to study projective 3-space and then merely generalize statements and proofs. As an example we consider the next lemma.

Lemma 3.9.7. *Any two planes of a projective 3-space S meet in a line.*

Proof. Let Π and Π' be distinct planes of S. Let p be a point of $\Pi \backslash \Pi'$ and let ℓ_1 and ℓ_2 be lines of Π on p. Π' is a hyperplane of S by corollary 2 to lemma 3.9.4. By lemma 3.9.6, the lines ℓ_1 and ℓ_2 meet Π'. The points of intersection must be distinct, and therefore generate a line (see exercise 43 of section 3.11) which is in both Π and Π'. □

This result can be generalized to give

Lemma 3.9.8. *Any two hyperplanes of a projective n-space meet in an $(n-2)$-dimensional subspace.*

We leave the proof of this as exercise 43.

The reader might have guessed by now that it is possible to construct projective spaces from vector spaces over skew-fields. This is indeed the case. Let V be an $(n+1)$-dimensional vector space over the skew-field F. The points of V are thus the $(n+1)$-tuples $(a_1, a_2, \ldots, a_{n+1})$ where $a_i \in F$, $1 \leq i \leq n+1$. Again, let the one-dimensional subspaces (lines on $(0,0,\ldots,0)$) and the two-dimensional subspaces (planes on $(0,0,\ldots,0)$) be the points and lines respectively of the new system $S(V)$.

Lemma 3.9.9. *$S(V)$ as defined above is a projective space.*

Proof. This follows easily from the fact that each plane of $S(V)$ is a projective plane by lemma 3.7.1. □

It is often useful when referring to projective n-space constructed from a finite vector space, to have a convenient and simple notation, indicating both the dimension and the order. We use $P(n, k)$ for such a space, the first entry, n, indicating the dimension, the second entry, k, indicating the order.

3.10 Desargues configurations again

We proved in theorem 3.7.4 that any projective plane $\Pi(V)$ constructed from a vector space V over a (skew-) field F is Desarguesian. What about other planes? It is possible to prove that any plane of a projective 3-space (and therefore of projective n-space, $n \geq 3$) is Desarguesian. (We do this below.) This leaves some puzzling questions. Are there projective planes which are not subspaces of larger projective spaces? *If* there are non-Desarguesian projective planes, then the answer to this must be yes.

To prove our claim above, we recall that in lemma 3.6.2 we proved that Π is Desarguesian if it has a complete set of central collineations. Thus to prove that a projective plane in projective 3-space is Desarguesian, it suffices to prove the next lemma.

Lemma 3.10.1. *Any plane Π of a projective 3-space S has a complete set of central collineations.*

Proof. Let c and ℓ be any point and line respectively of Π, and let x and x' be distinct points of Π, collinear with c, distinct from c, and not on ℓ. Let p be in $S\backslash\Pi$. Let Π' be the plane on p and ℓ. Let Π'' be the plane on p and the collinear points c, x and x'. Let $y \in xp\backslash\{x, p\}$ and choose $y' = cy \cap px'$. Note that y and y' are both in Π''.

We define a function f on points of Π as follows. *For* $z \in \Pi$, $f(z) = [(zy \cap \Pi')y'] \cap \Pi$, where $(zy \cap \Pi')y'$ is the line on y' and the point (by lemma 3.9.3) $zy \cap \Pi'$, and $f(z)$ is a point of Π by lemma 3.9.3. (See figure 3.10.1 which shows the elation case.)

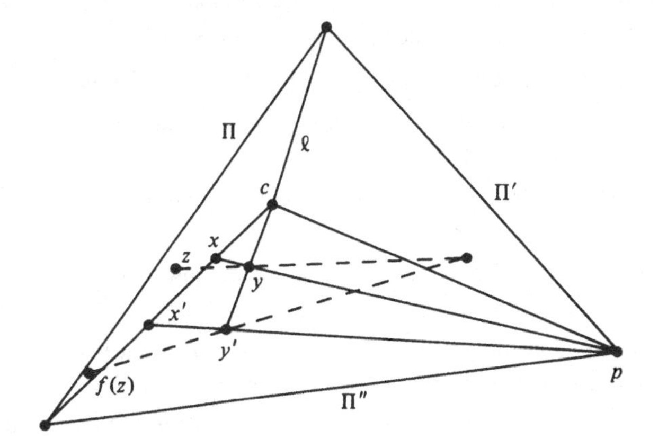

Figure 3.10.1.

If we define $g(z) = zy \cap \Pi'$, g having domain Π and range Π', and then define $h(w) = wy' \cap \Pi$, h having domain Π' and range Π, we can write

$f(z) = h \circ g(z)$. h and g are projections and therefore collineations (exercise 48 of section 3.11). Hence $f = h \circ g$ is a collineation. Moreover, f fixes c and each point of ℓ and maps x to x'. To show that f is a (c, ℓ)-collineation, it suffices to show that the points c, z and $f(z)$ above are collinear. But $\langle \{z, y, y'\} \rangle$ is a plane distinct from Π, which contains both c and $f(z)$. Since c, z and $f(z)$ are also in Π, these points are all in the line of intersection (lemma 3.9.8) of the two planes. □

Corollary. *Any plane of a projective n-space, $n \geq 3$, is Desarguesian.*

Proof. This follows from lemmas 3.10.1 and 3.6.2. □

There *are* non-Desarguesian projective planes. The smallest are of order 9. An example of a non-Desarguesian projective plane can be constructed as described in exercise 37 section 4.8.

3.11 Exercises

1. Show that the Fano plane is the smallest projective plane.
2. Prove that the duals of L1 and L2 hold in a projective plane.
3. A statement is *self-dual* if it is the same as its dual. Give an example of a self-dual statement.
4. Prove lemma 3.2.1 directly.
5. Prove that there is no projective plane on 29 points.
6. Show that a finite projective plane always has an odd number of points.
7. Label the projective plane of figure 3.1.1 in such a way that each line is of the form $\{x + a, y + a, z + a, w + a\}$ working modulo 13, where $0 \leq a \leq 12$ and $\{x, y, z, w\}$ is one of the lines. (See example 1.1.2.)
8. Draw a projective plane of order 4. (Hard: give the line sets.)
9. If Π' is a subplane of order m in a finite projective plane Π of order $n = m^2$, show that each point of Π lies on a line of Π'.
10. Find the possible orders of subplanes which may be contained in any projective plane of order 39.
11. Construct the projective plane of order 4 and show that it has a subplane of order 2.
12. Apply the embedding theorem, theorem 3.3.1, to the linear space of figure 3.3.1(*a*). Do you get the linear space of figure 3.3.1(*b*)?
13. Let Π be the Fano plane. Show that the completion of any quadrangle of Π is always Π. (See the final paragraphs of section 3.4.)
14. Find a central collineation of the projective plane of order 3.
15. Let Π be the extended real plane. Show that $f:(x, y) \rightarrow (-x, y)$ is a central collineation with f defined appropriately on the points at infinity. Find the centre and axis.

16. Find two non-identity (c, ℓ)-collineations of the projective plane of order 3, for fixed c and ℓ.

17. Prove that the set of all (c, ℓ)-collineations is a group.

18. Prove directly that an axial collineation is central.

19. A *correlation* of a projective plane is a 1–1 map from the set of all points and lines onto itself, such that a point is mapped to a line and a line to a point, and such that incidence is preserved (i.e., $p \in \ell$ if and only if $f(\ell) \in f(p)$). Find a correlation of the projective plane of order 2.

20. Show that the projective planes of orders 2 and 3 have complete sets of central collineations.

***21.** Show that the extended real plane has a complete set of central collineations.

22. Suppose Π is a finite projective plane of order k. Show that (i) if $c \in \ell$ and there are k distinct (c, ℓ)-elations, then Π is (c, ℓ)-transitive; (ii) if $c \notin \ell$ and there are $k-1$ distinct (c, ℓ)-homologies, then Π is (c, ℓ)-transitive.

23. Draw a Desargues configuration using a (c, ℓ)-elation.

24. We say that ℓ is a *translation line* of the projective plane Π if Π is (c, ℓ)-transitive for all $c \in \ell$. Prove that if Π is (p, ℓ)-transitive and (q, ℓ)-transitive for distinct points p and q of ℓ, then ℓ is a translation line of Π.

25. Explain why a field has at least two elements.

26. Check that GF(7) is a field.

27. Let F be the field of integers modulo 5. Show that $f(x) = x^2 + 1$ is reducible over F and that $g(x) = x^2 + 2$ is irreducible over F. Let a be a root of $g(x)$. List the 25 elements of $F(a)$ and check that they form a field.

28. Find a field F and use it to construct the projective plane of order 3.

29. Find in $P(2, 11)$ the intersection of the lines $\langle 1, 2, 3 \rangle$ and $\langle 2, 4, 5 \rangle$.

30. In a vector space V over GF(k), k finite, show that there are $k + 1$ lines on the origin in any plane through the origin.

31. Show directly that $P(2, k)$ has $k^2 + k + 1$ points.

32. Prove that any two projective planes constructed from the same field F are isomorphic.

33. If V is a three-dimensional vector space over GF(q), show that there are

$$\frac{(q^3 - 1)(q^3 - q)}{(q^2 - 1)(q^2 - q)} = q^2 + q + 1$$

two-dimensional subspaces.

34. Prove that if $[x, y, z]$ and $[x', y', z']$ are distinct points of $\Pi(V)$ over a field, then $\langle yz' - y'z, zx' - z'x, xy' - x'y \rangle$ is a line (i.e., not all three co-ordinates are zero), and prove algebraically that it is the line on the two given points.

35. Draw the three-dimensional vector space over GF(3).

36. In $P(2, 5)$ choose c, x, x', y, y', z, z' as in a Desargues configuration. Calculate u, v and w and show that they are collinear.

37. Check that all the results of section 3.7 hold without using the commutative property $ab = ba$ of the field.

38. Find the number of lines in $P(3, k)$.

39. How many planes are there on a given line in $P(3, k)$?

***40.** Derive a theory of duality in projective 3-space. (See section 3.1.) Can you generalize to n-space?

41. How many hyperplanes are there in $P(n, k)$?

42. If S is a three-dimensional projective space containing disjoint lines ℓ and ℓ' and p is a point in neither ℓ nor ℓ', show that there is exactly one line in S which contains p and meets both ℓ and ℓ'.

43. Prove that any two hyperplanes of projective n-space meet in an $(n-2)$-dimensional subspace.

44. Show that $P(n, k)$ has the exchange property.

45. Show that two distinct planes in projective n-space cannot intersect in a plane.

46. Prove that if S is a projective space with dimension at least 3 in which any two planes intersect in a line, then S has dimension 3.

47. Using the construction of lemma 3.9.9, construct a projective 3-space using GF(2), and then using GF(3).

48. Let Π and Π' be projective planes in a projective 3-space S and let $p \in S \backslash (\Pi \cup \Pi')$. For all $x \in \Pi$ define $g(x) = xp \cap \Pi'$. The map g is called a *projection of* Π onto Π'. Show that g is an isomorphism.

49. Draw the figure for lemma 3.10.1 in the homology case.

50. Show that a projective 3-space with planes as 'blocks' does not form a Steiner system.

***51.** Prove that if a projective plane Π is Desarguesian, then it has a complete set of central collineations.

52. Show that a plane of a projective space meets any hyperplane either in a line or in the whole plane.

53. Let L be a finite linear space in which no line has just two points. Let p, q, r. s be any quadrangle (set of four points, no three collinear) of L. Suppose that whenever the lines pq and rs intersect, then so do the lines pr and qs. Prove that L is a projective space.

4

Affine spaces

But if you look into the depths of your own consciousness – assuming such depths exist – you will find, I believe, an eternal distinction maintained, in this respect, between straight and curved Lines: so that Lines of the one kind *must*, if they approach, ultimately meet, whereas those of the other kind need not.

Lewis Carroll *Euclid and his Modern Rivals*

As in chapter 3, we shall restrict ourselves to the planar case for the greater part of this chapter.

4.1 Affine planes

An *affine plane* is a linear space A with the properties

A1 any point p not on a line ℓ is on precisely one line missing ℓ, and

A2 there exists a set of three non-collinear points.

If ℓ and ℓ' are lines of an affine plane and $\ell = \ell'$ or ℓ and ℓ' do not meet, we say they are *parallel*, and shall write $\ell \| \ell'$.

Example 4.1.1. Ordinary real 2-space is an example of an affine plane.

Example 4.1.2. The linear space of figure 4.1.1 is an affine plane.

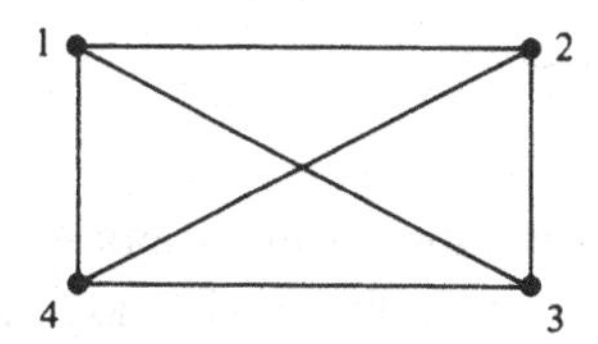

Figure 4.1.1.

Notice that the lines $\{1,3\}$ and $\{2,4\}$ are parallel.

Example 4.1.3. The linear space of figure 4.1.2 is also an affine plane. We saw this linear space as figure 2.1.2 and examined some of its characteristics.

Lemma 4.1.1. *If S is a linear space with more than one line, and if $c_{ij} = b_i - 1$ for each $p_i \notin \ell_j$, then S is an affine plane.*

Proof. It is easy to see that A1 and A2 are immediately satisfied. □

In exercise 12 of section 2.7, we introduced the concept of an equivalence relation, and showed that parallelism in the real plane was such a relation. In fact, this is true in *any* affine plane, as we show in the next lemma.

Lemma 4.1.2. *Parallelism is an equivalence relation on the set of lines in an affine plane.*

Proof. Any line is parallel to itself by definition, so that the reflexive property is automatically true. It is also clear that if ℓ and ℓ' do not meet, then ℓ' and ℓ do not meet; so symmetry is also trivial. To show the transitive property, let ℓ and ℓ' be parallel, and let ℓ' and ℓ'' be parallel, but suppose ℓ and ℓ'' meet in a point p, say. Then p is on two lines which miss ℓ' and we contradict A1. So transitivity also holds. □

The equivalence classes under the equivalence relation of lemma 4.1.2 are known as *parallel classes.*

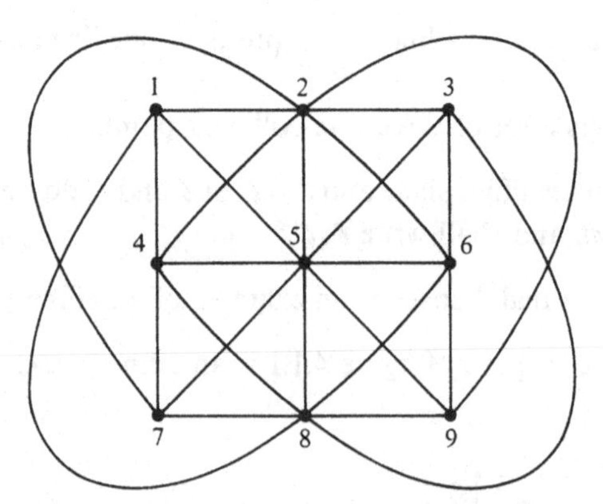

Figure 4.1.2.

The term 'affine plane' suggests that an affine plane is indeed a plane, that is, has dimension 2. We prove this now in the special case that each line of an affine plane has at least three points.

Lemma 4.1.3. *An affine plane A in which every line has at least three points has dimension 2.*

Proof. We show that any set of three non-collinear points generates A. (Since A *has* such a set of points by A2, we may conclude that A has dimension 2.) Let ℓ be any line (the closure of two points) and p a point not on ℓ. Then the closure of ℓ and p contains the points of all lines on p

except possibly the points of the line $\hbar$ on p missing ℓ. (See figure 4.1.3.)

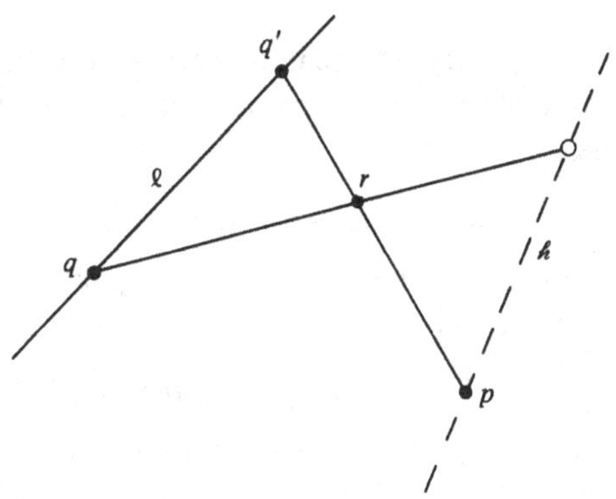

Figure 4.1.3.

Let $q, q' \in \ell$. As each line has at least three points, there is a point r on $pq' \setminus \{p, q'\}$, and the line qr is a subset of $\langle \{p\} \cup \ell \rangle$. But qr meets $\hbar$ in a point different from p, and so $\hbar \subseteq \langle \{p\} \cup \ell \rangle$. Thus *all* points are in $\langle \{p\} \cup \ell \rangle$. □

What happens if some lines have only two points? We have an example in figure 4.1.1. This space has dimension 3, as the reader can check. We shall see later that this is the *only* affine plane which is not a plane!

We saw in section 3.1 (lemma 3.1.5) that the dual of the axioms for a projective plane hold in a projective plane. This is not true for an affine plane. The dual of 'two points are on a line' is 'two lines meet in a point', and this is not always true by A1.

4.2 Finite affine planes

Suppose in this section that A is an affine plane with a finite number of points. We shall compute the total number of points, the total number of lines, the number of lines per point and points per line, etc.

Lemma 4.2.1. *All lines in A have the same number of points, and all points are on the same number of lines.*

Proof. Let ℓ and ℓ' be distinct lines. We shall show $v(\ell) = v(\ell')$. If all points are on either ℓ or ℓ', then $\ell \parallel \ell'$ by A1. If ℓ or ℓ' – say ℓ – has at least three points, let p, q and r be three of these, and let $p' \in \ell'$. Then p' is on the lines $p'p$ and $p'q$ both of which miss $p'r$, contradicting A1. Thus ℓ and ℓ' each have two points and A is the affine plane of figure 4.1.1.

We may now suppose that some point p is on neither ℓ nor ℓ'. By A1, $v(l) = b(p) - 1$ and $v(\ell') = b(p) - 1$.

It follows immediately that each point is on $v(\ell)+1$ lines for any line ℓ. □

Letting k be the number of points per line in A, we call k the *order* of A. (Compare with the definition of order in a *projective* plane.) It follows that any point is on $k+1$ lines.

Lemma 4.2.2. *If A has order k then A has k^2 points.*

Proof. Fix a point p. As p is on $k+1$ lines each with k points, we count $(k+1)(k-1)-1=k^2$ points in all. □

Lemma 4.2.3. *If A has order k then each line is parallel to k lines.*

Proof. Let ℓ be any line and ℓ' any line meeting ℓ in one point. Each line h parallel to ℓ meets ℓ', as otherwise there are two lines on $\ell \cap \ell'$ parallel to h, contradicting A1. By A1, each point on ℓ' is on a unique line parallel to ℓ, and since ℓ' has order k, it follows that there are k lines in all parallel to ℓ. □

Lemma 4.2.4. *If A has order k, then A has k^2+k lines.*

Proof. We count the total number of lines b, by fixing a line ℓ and counting the number of lines which meet it in a single point and the number of lines parallel to it. By lemma 4.2.3, k lines are parallel to ℓ. By lemma 4.2.1, each point of ℓ is on k lines if we exclude ℓ itself. So $k \cdot k$ lines meet ℓ in one point. Thus $b=k^2+k$. □

Corollary 1. *If A has order k then each line ℓ meets k^2 other lines.*

Corollary 2. *If A has order k then there are $k+1$ parallel classes.*

It follows from lemmas 4.1.3 and 4.2.1 that the affine plane of order 2 is the unique affine plane which is not a plane. We use the word 'unique' here, and the reader should check that there *is* only one affine plane of order 2.

4.3 Embedding an affine plane in a projective plane

There is an intimate relationship between affine and projective planes. We plan to uncover this relationship in this section. Before reading the proof of theorem 4.3.1, we advise the reader to check the definition of the term 'embedding' (section 3.3).

Theorem 4.3.1. *Any affine plane A is embeddable in a projective plane Π so that the points of A are the points of $\Pi \backslash \ell$ for some line ℓ of Π. If A has order k then Π also has order k.*

Proof. For any line ℓ, let $[\ell]$ denote the parallel class containing ℓ. By lemma 4.1.2 (parallelism is an equivalence relation), $\ell' \in [\ell]$ if and only if

$[\ell'] = [\ell]$. Construct the system $\Pi = (P', L')$ from $A = (P, L)$:

$$P' = P \cup \{[\ell] \mid \ell \in L\},$$
$$L = \{\{\ell \cup [\ell] \mid \ell \in L\}, \{[\ell] \mid \ell \in L\}\}.$$

In other words, the points of P' are the points of P plus some extra points corresponding to the sets of lines $[\ell]$. The lines of L' are made up of the lines of L, each receiving an additional 'point', along with one extra line, the line consisting entirely of the new points.

It is easy to see that there are at least two parallel classes in any affine plane (see lemma 4.2.3), and so all lines of L' have at least two points. Any two distinct lines ℓ and ℓ' of L which were parallel in L now meet in $[\ell] = [\ell']$. Moreover, each line meets the new line $\{[\ell] \mid \ell \in L\} = \ell_\infty$. So PP1 is satisfied. It is easy to check that PP2 is also true.

Thus $\Pi = A \cup \ell_\infty$ and, if A has order k, so has Π. We leave it to the reader to give a precise description of the embedding function. □

This construction should be reminiscent of the construction of example 2.1.3 where we embedded real 2-space in a projective plane.

Theorem 4.3.2. *Let Π be any projective plane and ℓ a line of Π. Then $\Pi \setminus \ell$ is an affine plane.*

Proof. It is clear that A2 holds. To show A1, let ℓ' be any line of $\Pi \setminus \ell$ and p a point not on ℓ. If $\ell \cap \ell' = q$ in Π, then the unique line on p missing ℓ in $\Pi \setminus \ell$ is $pq \setminus \{q\}$ (which is still a line since lines in Π each have at least three points). □

Because of the connection between affine and projective planes which we have just established, it follows that the existence results for projective planes can be applied to affine planes (see section 3.2). Thus there are affine planes of orders 2, 3, 4, 5, 7, 8, 9. There is none of order 6 or 10.

The reader should also notice that all of the counting results of section 4.2 follow as a corollary from theorem 4.3.1.

We pose an interesting question (without giving an answer) at the end of this section: take an affine plane A and embed it by the method of theorem 4.3.1 in a projective plane Π, in such a way that $\Pi = A \cup \ell$. Let $\ell' \in \Pi$, $\ell' \neq \ell$. Obtain the affine plane A' from $\Pi \setminus \ell'$ by the method of theorem 4.3.2. Is A isomorphic to A'?

4.4 Collineations in affine planes

The reader should check the definition of collineation before continuing.

Example 4.4.1. Let A be the affine plane on four points. Label the points as in figure 4.1.1. Define $f(1)=4$, $f(2)=3$, $f(4)=1$ and $f(3)=2$. It is easy to check that this is a collineation, as is g defined by $g(1)=1, g(2)=3, g(3)=4, g(4)=2$. Of course, the identity is always a collineation.

We leave the reader to find non-identity collineations of the affine planes of orders 3 and 4.

Example 4.4.2. Let A be real 2-space. Define $f((x,y))=(ax,ay)$, $a\neq 0$, a real, for each point (x,y). The origin maps to itself, and it is the only point which does so if $a\neq 1$. Any line through the origin maps to itself since the equation $y=mx$ becomes $ay=max$, and $x=0$ becomes $ax=0$. A general line of the form $y=mx+b$ becomes $y=mx+b/a$ so that lines not on $(0,0)$ are not fixed but map to lines with the same slope.

Such a collineation is called a *dilatation.* (We saw its counterpart in the extended real plane in example 3.5.2.) Points are either 'pulled away from' or 'drawn towards' $(0,0)$.

Example 4.4.3. Let A be real 2-space. Define $f((x,y))=(x+a,y+a)$, $a\neq 0$. Clearly no point of A is fixed. Lines of the form $y=mx+b$ become lines of the form $y+a=m(x+a)+b$, or $y=mx+(b+(m-1)a)$, so that slopes are preserved. Clearly vertical lines map to vertical lines. The function f is known as a *translation.*

Example 4.4.4. Again let A be real 2-space. This time define $f((x,y))=(-x,y)$. Each point of the line $x=0$ is fixed. Lines of the form $y=mx+b$ go to lines of the form $y=-mx+b$, so that slope is not preserved. This collineation is known as a *reflection* as it mirrors each point through the y-axis. Clearly, the function mapping each point (x,y) to the point $(x,-y)$ should also be labelled a reflection.

Example 4.4.5. As a last example in real 2-space define $f((x,y))=(x/\sqrt{2}+y/\sqrt{2}, x/\sqrt{2}-y/\sqrt{2})$. The distance between (x,y) and the origin is $\sqrt{(x^2+y^2)}$. The distance between $f((x,y))$ and the origin is $\sqrt{((x/\sqrt{2}+y/\sqrt{2})^2+(x/\sqrt{2}-y/\sqrt{2})^2)}=\sqrt{(x^2+y^2)}$. The line $y=mx$ through the origin becomes the line $y=((1-m)/(1+m))x, m\neq -1$. Thus f is a *rotation.*

Many other collineations of the real plane can be defined using combinations of the collineations of examples 4.4.2, 4.4.3, 4.4.4 and 4.4.5.

We now turn to the general theory of collineations in affine planes.

Lemma 4.4.1. *Let f be a collineation of A, and ℓ and ℓ' be distinct lines. Then $\ell \parallel \ell'$ if and only if $f(\ell)\parallel f(\ell')$.*

Proof. We note first of all that, since f is a collineation, $f(\ell)$ and $f(\ell')$ are distinct lines.

Let $\ell \parallel \ell'$. If $p\in f(\ell)\cap f(\ell')$ then there are (distinct) points q and q'

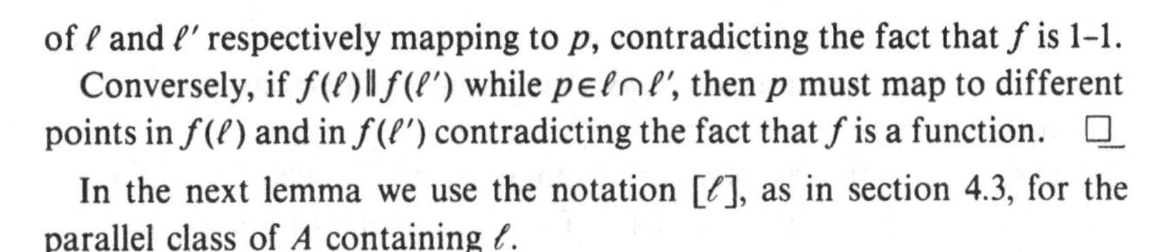

of ℓ and ℓ' respectively mapping to p, contradicting the fact that f is 1–1.

Conversely, if $f(\ell) \| f(\ell')$ while $p \in \ell \cap \ell'$, then p must map to different points in $f(\ell)$ and in $f(\ell')$ contradicting the fact that f is a function. □

In the next lemma we use the notation $[\ell]$, as in section 4.3, for the parallel class of A containing ℓ.

Lemma 4.4.2. *For any collineation f of A, $f([\ell]) = [f(\ell)]$.*

Proof. It follows from lemma 4.4.1 that a whole parallel class gets mapped onto a whole parallel class. Clearly, since ℓ gets mapped to $f(\ell)$, the parallel class $[\ell]$ gets mapped to the parallel class $[f(\ell)]$. □

A *dilatation* of A is a collineation f such that $\ell \| f(\ell)$ for each line ℓ.

This generalizes the concept of dilatation in R^2, but *be careful*: translations in R^2 are also included in the above definition.

Lemma 4.4.3. *For any dilatation f of A, $f([\ell]) = [\ell]$.*

Proof. This follows immediately from lemma 4.4.2 and the definition of dilatation. □

We shall now break the class of dilatations down into two types, corresponding to 'dilatations' and 'translations' in R^2. Our apologies for the confusing mixture of definitions, but the terminology is reasonably standard. The reader is advised to concentrate on the general definitions in an arbitrary affine plane.

A dilatation with a fixed point is called a *homothety*, and a fixed point is called a *centre*.

Lemma 4.4.4. *A homothety f satisfies $f(\ell) = \ell$ for each line ℓ on a centre.*

Proof. As $f(\ell) \| \ell$ and the point c on ℓ is fixed, $f(\ell) = \ell$. □

The map of example 4.4.1 is clearly not a homothety, while the map of example 4.4.2 is. The reader should verify that the affine plane of order 2 has no non-identity homothety (exercise 24 of section 4.8), and also that the affine plane of order 3 does have them (exercise 23 of section 4.8).

Lemma 4.4.5. *A non-identity homothety has precisely one fixed point* (*the centre*).

Proof. Suppose c and c' are distinct fixed points of the homothety f. Let x be any point not on cc'. Then $f(xc) = xc$ and $f(xc') = xc'$ by lemma 4.4.4. So $f(x) \in xc \cap xc' = \{x\}$. Hence each point not on cc' is fixed. Now if $x \in cc' \setminus \{c, c'\}$, choose $y \notin cc'$. So c and y are fixed and $x \notin cy$. Arguing as above shows again that x is fixed. So f is the identify. □

A *translation* is either the identity map or a dilatation with no fixed points.

Notice that the set of dilatations is the union of the set of homotheties and the set of translations, the only element in the intersection being the identity.

Translations of the real plane are examples of more general 'translations' in an affine plane. (See example 4.4.3.) The example of 4.4.1 is a translation. The map $f{:}1 \to 4, 2 \to 5, 3 \to 6, 4 \to 7, 5 \to 8, 6 \to 9, 7 \to 1, 8 \to 2, 9 \to 3$ of the affine plane in figure 4.1.2 is a translation.

Lemma 4.4.6. *Suppose f is a non-identity translation of A. Then, for any point p, $f(pf(p)) = pf(p)$; that is, f fixes the line $pf(p)$. Moreover, the set of all lines fixed by f forms a parallel class.*

Proof. Since f is a dilatation, $f(pf(p)) \parallel pf(p)$. But both lines contain $f(p)$, and so they are the same line.

If distinct lines ℓ and ℓ' are fixed by f, and p is in their intersection, then $f(p) \neq p$ while $f(p)$ must be in both ℓ and ℓ', which is a contradiction. So any two lines fixed by f are parallel to each other. Moreover, as *every* point p is on a line which is fixed, the set of lines of the form $pf(p)$, fixed by f, forms a parallel class. □

The centre of a homothety in A acts like the centre of a central collineation of a projective plane, apart from the fact that we have no axis. A translation also acts like a central collineation in many ways, but one which has lost its centre, thus leaving a parallel class of lines preserved by the function.

Suppose we embed the affine plane A in the projective plane Π using the method of section 4.3, so that $\Pi = A \cup \ell_\infty$. Let f be a dilatation of A. As f maps parallel classes to parallel classes (lemma 4.4.2), we see how to define an extension f_Π of f on ℓ_∞. That is, the function f_Π on Π acts exactly as f does on A, and takes each point $[\ell]$ of ℓ_∞ to the point $[f(\ell)]$ on ℓ_∞. This produces a collineation of Π. But if f is a dilatation, $[\ell] = [f(\ell)]$ by lemma 4.4.3, so that f_Π fixes each point of ℓ_∞. Thus f_Π has an *axis* and so also a *centre*.

The map f_Π constructed from the dilatation f is thus a central collineation. If f is a homothety, the centre is in A and not on ℓ_∞, so that f_Π is a homology. If f is a translation, the centre c must be on ℓ_∞ so that f_Π is an elation.

Suppose we were to start with a projective plane Π and a central collineation f_Π. If we were to remove the axis, it is not difficult to see that f_Π reduces to a dilatation on A, which will be a homothety or a translation depending on whether f is a homology or an elation.

But why remove the axis of f_Π from Π? Is it possible to remove some other line and still have f_Π reduce to a function with interesting properties?

Indeed it is possible, and to introduce such a function (which must fix every point of a line of A), we make the following definition.

A collineation of A which fixes each point of a line of A is called an *affine perspectivity*. The fixed line is called the *axis*.

The reflection in the y-axis of the real plane in example 4.4.4 is an affine perspectivity in which the y-axis is the axis.

Example 4.4.6. For the affine plane of figure 4.1.2, the map $f:1\to 1, 2\to 2, 3\to 3, 4\to 5, 5\to 6, 6\to 4, 7\to 9, 9\to 8, 8\to 7$ is an affine perspectivity.

Lemma 4.4.7. *A non-identity affine perspectivity f of A has no fixed points not on the axis, and hence has precisely one axis.*

Proof. Let the axis be ℓ and suppose that $f(p)=p$ for some point $p\notin\ell$. Hence, any line on p which meets ℓ is fixed by f as it has two fixed points. It follows that the line on p parallel to l can only map to itself.

Since f is not the identity, for some point x, $f(x)\neq x$. Then $f(x)\in f(px)=px$ since all lines on p are mapped to themselves. Choose $q\in\ell\backslash px$. Let h be the line on p parallel to qx. Again, h on p implies $f(h)=h$, while $f(qx)=qf(x)\neq qx$. Finally, lemma 4.4.1 tells us that $h\parallel qx$ implies $f(h)\parallel qf(x)$ which is not true. □

For an affine perspectivity f, we define a *trace* of f to be a line of the form $pf(p)$ where $p\neq f(p)$.

We saw in lemma 4.4.6 that, for translations, such lines play special roles. This is also true for affine perspectivities as we see in the next three lemmas.

Lemma 4.4.8. *An affine perspectivity fixes traces.*

Proof. We may assume that f is a non-identity affine perspectivity with axis ℓ, and that p is a point such that $p\neq f(p)$.

Suppose $pf(p)\cap\ell=q$. (Clearly $q\neq p, f(p)$.) Then $f(pf(p))=f(pq)=f(p)q=pq$.

Suppose $pf(p)\parallel\ell$. By lemma 4.4.1, this implies $f(pf(p))\parallel f(\ell)$. That is, $f(pf(p))\parallel\ell$. By A1, and the fact that $f(p)$ is on both $pf(p)$ and $f(pf(p))$, $pf(p)=f(pf(p))$. □

Lemma 4.4.9. *If f is a non-identity affine perspectivity such that $f(\ell)=\ell$ and ℓ is not the axis of f, then ℓ is a trace.*

Proof. This is trivial, since for any point p on such a line ℓ, $f(p)\in\ell$ also. □

Lemma 4.4.10. *Let f be a non-identity affine perspectivity with axis ℓ. If some trace meets ℓ then the set of traces forms a parallel class. If no trace meets ℓ then the set of traces together with ℓ forms a parallel class.*

Proof. Let ℓ be the axis of f and let ℓ' be a trace. Suppose ℓ' meets ℓ. Let ℓ'' be any line parallel to ℓ'; so ℓ'' meets ℓ in a point p say. Now $\ell' \| \ell''$ implies $f(\ell') \| f(\ell'')$ by lemma 4.4.1. That is, $\ell' \| f(\ell'')$. But $p \in f(\ell'')$ so that $f(\ell'')$ must be ℓ''. We have proved that if ℓ' meets ℓ then every line parallel to ℓ' is a trace. Are all traces parallel to ℓ'?

Let ℓ'' be any trace. Write $\ell'' = qf(q)$. Let h be the unique line on q parallel to ℓ'. By the argument of the preceding paragraph, h is a trace. Therefore $h = qf(q) = \ell''$.

Suppose now that the trace ℓ' does not meet ℓ. Let ℓ'' be a trace distinct from ℓ'. If $\ell' \cap \ell'' = p$, a point, then clearly p is fixed, and so on ℓ by lemma 4.4.7, a contradiction. So any trace is parallel to ℓ'. Let p be any point not on ℓ. Then p is on a trace $pf(p)$ as, again by lemma 4.4.7, $p \neq f(p)$. Hence the traces form the parallel class $[\ell']$. □

Once again, embed A in Π via the method of section 4.3 and let $\Pi = A \cup \ell_\infty$. If f is a (non-identity) affine perspectivity of A, we need to define an extension f_Π of f in Π. Clearly, $f_\Pi([\ell]) = [f(\ell)]$, using lemma 4.4.2, gives us the extension. The set of traces determines a fixed point of ℓ_∞. If the traces are parallel to the axis ℓ of f, then this fixed point is on ℓ_∞ and we have an axis of an elation of Π. If the traces meet ℓ, we have a homology. (The reader should check this in detail.)

If Π is a projective plane with a central collineation f, it is not always the case that any affine plane $A = \Pi \backslash \ell$ (theorem 4.3.2) has either a dilatation or an affine perspectivity induced on it via f. The reader should check to see what the other possibilities are. In particular, we refer to exercise 28 of section 4.8.

The diagram below explains what happens when one takes a projective plane Π with a central collineation f and removes particular lines.

	Π		A
Remove the axis	homology	→	homothety
	elation	→	translation
Remove a line, not the axis ℓ, on the centre	homology	→	affine perspectivity with traces meeting ℓ
	elation	→	affine perspectivity with traces parallel to ℓ

The property of having a complete set of central collineations is an important one in projective planes, since we saw in section 3.6 that it is equivalent to the property of being Desarguesian. We want to look at the equivalent notions in affine planes.

An affine plane A is said to have a *complete set of dilatations* if, given any distinct collinear points c, x and x' there is a homothety f of A with centre c such that $f(x)=x'$, *and* given any points p and q of A there is a translation of A mapping p to q.

Note that if $p=q$ in the above definition, the translation chosen must be the identity.

An affine plane A is said to have a *complete set of affine perspectivities* if, given any line ℓ and points p and q not on ℓ, there is an affine perspectivity f of A with axis ℓ such that $f(p)=q$.

4.5 The Desargues configuration in affine planes

An affine plane A is said to be *Desarguesian* if and only if the projective plane $\Pi = A \cup \ell_\infty$ is Desarguesian.

The next result follows immediately.

Lemma 4.5.1. *If A is a Desarguesian affine plane, then for any set of distinct points c, x, x', y, y', z, z' such that c, x, x' and c, y, y' and c, z, z' are collinear and such that x, y, z and x', y', z' are triangles,*

(a) *if all points $xy \cap x'y'$, $xz \cap x'z'$, $yz \cap y'z'$ exist, then they are collinear,*
(b) *if $xy \parallel x'y'$ and $xz \parallel x'z'$, then $yz \parallel y'z'$,*
(c) if $u = xy \cap x'y'$ *and* $v = xz \cap x'z'$ *exist, but* $yz \parallel y'z'$, *then* $uv \parallel yz$.

Moreover, for any set of distinct points x, x', y, y', z, z' such that $xx' \parallel y'y' \parallel zz'$ and such that x, y, z and x', y', z' are triangles,

(d) *if all points $xy \cap x'y'$, $xz \cap x'z'$, $yz \cap y'z'$ exist, then they are collinear,*
(e) *if $xy \parallel x'y'$ and $xz \parallel x'z'$, then $yz \parallel y'z'$,*
(f) *if $u = xy \cap x'y'$ and $v = xz \cap x'z'$ exist, but $yz \parallel y'z'$, then $uv \parallel yz$.*

We now show that the converse of this lemma also holds.

Lemma 4.5.2. *Let A be an affine plane in which (a)–(f) of lemma 4.5.1 hold. Then A is Desarguesian.*

Proof. It suffices to show that $\Pi = A \cup \ell_\infty$ is Desarguesian. We break the proof into eight cases. Let c, x, x', y, y', z, z', u, v and w be as in a Desargues configuration in Π. We consider (i) none of the ten points is on ℓ_∞, (ii) $u, v \in \ell_\infty$, (iii) $c \in \ell_\infty$, but no other point, (iv) $c, x, x' \in \ell_\infty$, (v) $x, y, u \in \ell_\infty$, (vi) $u \in \ell_\infty$, but no other point except possibly c, (viii) $x \in \ell_\infty$, but no other point except possibly w, (viii) $x, y' \in \ell_\infty$.

The cases (i) and (iii) follow directly from (a) and (b) respectively. Case (ii) follows from (b) and (e). Suppose as in (iv) that $c, x, x' \in \ell_\infty$. Let $\bar{u} = vw \cap xy$ and let $w = \bar{c}$. We consider a new system of points $\bar{c}, z, y, v, \bar{u}, z', y'$ (see figure 4.5.1) where $\bar{c}, z, y$ and $\bar{c}, v, \bar{u}$ and $\bar{c}, z', y'$ are collinear, and it is

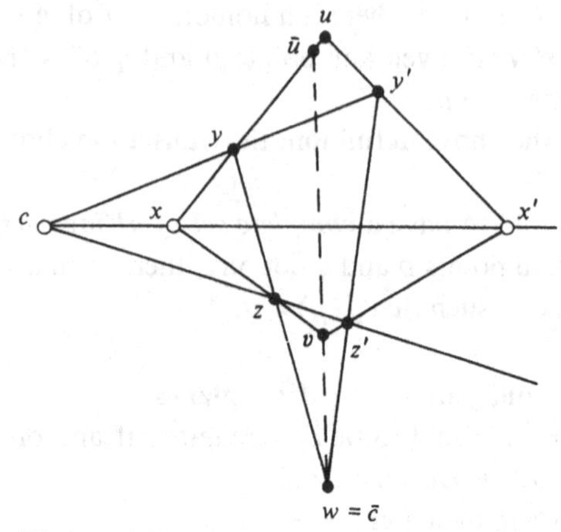

Figure 4.5.1.

easy to check that z, z', v and $y, y', \bar{u}$ are indeed triangles. In A, $zv \parallel y\bar{u}$ and $zz' \parallel yy'$, so that by (b), $\bar{u}y' \parallel vz'$. As x' is the only point of vz' on ℓ_∞ it must be the case that $x' \in \bar{u}y'$. But $x' \in uy'$, and we are forced to conclude that $u = \bar{u}$. Hence u, v and w are collinear.

Suppose as in (v) that $x, y, u \in \ell_\infty$. Let $z' = \bar{c}$ and consider the system $\bar{c}, c, z, x', v, y', w$, where $\bar{c}, c, z$ and $\bar{c}, x', v$ and $\bar{c}, y', w$ are collinear and c, x', y' and z, v, w are triangles. Then $cx' \cap zv = x$ and $cy' \cap zw = y$. By (b), x, y and $x'y' \cap vw$ are collinear. But $u \in xy$ and $u \in x'y'$. Therefore $u \in vw$ and we are done.

Case (vi) follows directly from (c) and (f). Suppose as in (vii) that $x \in \ell_\infty$, but no other point except possibly w. Consider the system x, c, x', z, v, y, u where x, c, x' and x, z, v and x, y, u are collinear, and c, y, z and x', u, v are triangles. Let $\bar{w} = yz \cap vu$. If $\bar{w} \notin \ell_\infty$, then by (d), y', z' and $\bar{w}$ are collinear, forcing $w = \bar{w}$ so that u, v and w are collinear. If $\bar{w} \in \ell_\infty$ we get the same conclusion using (vi).

We leave the proof of case (viii) as an exercise for the reader. □

For an example of a non-Desarguesian affine plane and hence also a non-Desarguesian projective plane, we refer the reader to exercise 37 of section 4.8, and also to Hilbert (1962).

Lemma 4.5.3. *A projective plane Π has a complete set of central collineations if and only if for any line ℓ, the affine plane $A = \Pi \setminus \ell$ has a complete set of dilatations and of affine perspectivities.*

Proof. Suppose Π has a complete set of central collineations. Let c, x and x' be distinct collinear points of A. Then there is a central collineation of Π with centre c, axis ℓ, which maps x to x'. In A, this map reduces to a homothety with centre c mapping x to x'.

Let p and q be distinct points of A. Then there is a central collineation of Π with centre $c = pq \cap \ell$, axis ℓ, mapping p to q. In A this reduces to a translation.

Finally, choose any line ℓ' of A and distinct points p and q not on ℓ'. There is a central collineation of Π with axis ℓ', centre $c = pq \cap \ell$ which maps p to q. This reduces to an affine perspectivity in A.

Now suppose that A has a complete set of dilatations and of affine perspectivities. The fact that Π has a complete set of central collineations follows from the remarks made in the previous section concerning extending functions in A to functions in Π. We leave the details as an exercise. □

4.6 Co-ordinatization in affine planes and the Pappus configuration

In section 3.7 we saw how to use skew-fields to co-ordinatize projective planes. They can also be used to co-ordinatize affine planes. In fact the method of co-ordinatization is much simpler in the latter case, and is similar to the co-ordinatization of the Euclidean plane.

Let F be a skew-field and define

$$A(F) = \{(x, y) \mid x, y \in F\}.$$

$A(F)$ will be the set of *points* of the affine plane we want to construct.

A *line* of $A(F)$ is the set of points (x, y) satisfying an equation of the form

$$y = mx + b$$

or of the form

$$x = b$$

where m and b are fixed elements of F. As in Euclidean geometry, we shall call m the *slope of the line* in the former case.

Lemma 4.6.1. *The system $A(F)$, with lines as defined above, is an affine plane.*

Proof. The proof of the necessary axioms is identical to that for the Euclidean plane, and we leave it as an exercise for the reader. □

Example 4.6.1. Let $F = \mathrm{GF}(3)$. $A(F)$ then has nine points and the equations of the lines are

$$x=0,\ x=1,\ x=2,\ y=0,\ y=1,\ y=2,$$
$$y=x,\ y=x+1,\ y=x+2,\ y=2x,\ y=2x+1,\ y=2x+2.$$

The plane is drawn with appropriate co-ordinates in figure 4.6.1.

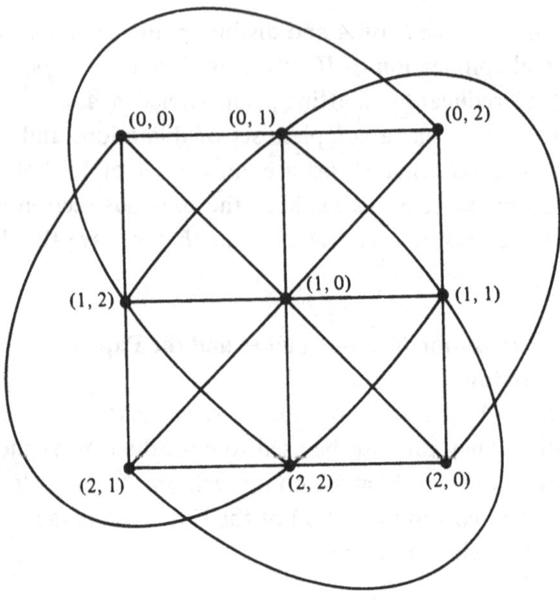

Figure 4.6.1.

As in the Euclidean plane, the parallel classes of lines are determined by the slopes of the lines. We prove this below. Let us say that any line of the form $x=b$ has 'infinite slope'.

Lemma 4.6.2. *The set of all lines of the same slope forms a parallel class in A.*

Proof. Any two lines with infinite slope obviously do not meet. Clearly, any line of the form $x=b$ meets any line of the form $y=mx+b'$ since we can solve for y in the second equation upon setting $x=b$. So all lines with infinite slope form a parallel class.

If $\ell: y=mx+b$ and $\ell': y=mx+b'$ intersect in a point (x_1, y_1) say, then subtracting equations gives $b=b'$. So any two lines with the same slope are parallel. Suppose ℓ has equation $y=mx+b$, ℓ' has equation $y=m'x+b'$ and $\ell \| \ell'$. Then $\ell' \| \ell''$ where $\ell'': y=m'x+b$, and so $\ell \| \ell''$. However, $(0, b)$ is a point of both ℓ and ℓ'', and so $\ell=\ell''$. Thus ℓ and ℓ' have the same slope. □

The obvious question now is whether there is any connection between this co-ordinatization for affine planes, and the one we used for projective planes. We answer this in the affirmative via the next lemma.

Lemma 4.6.3. *Let F be a skew-field, and let $\Pi(A(F))$ be the projective plane obtained as in section 4.3 from the affine plane $A(F)$. Then $\Pi(A(F))$ is isomorphic to the projective plane $\Pi(V)$ obtained from a three-dimensional vector space V over F.*

Proof. To demonstrate the isomorphism, we set up as usual a 1–1 onto function between the two systems, and then show that we have a linear function.

Recall that the points of the projective plane obtained from a vector space have co-ordinates of the form $[x, y, z]$ where such a point is identified with $[kx, ky, kz]$ for any non-zero skew-field element k.

For $(x, y) \in A$ define

$$f((x, y)) = [x, y, 1].$$

Any point on the line at infinity ℓ_∞ of $\Pi(A(F))$ can be identified with the corresponding parallel class of lines, each of which has the same slope by lemma 4.6.2.

For $p \in \ell_\infty$ corresponding to the parallel class with slope m, define

$$f(p) = [1, m, 0].$$

For $p \in \ell_\infty$ corresponding to the parallel class with infinite slope, define

$$f(p) = [0, 1, 0].$$

It is easy to check that f is a 1–1 and onto map.

The points on ℓ_∞ of $\Pi(A(F))$ correspond precisely to the points of $\Pi(V)$ with third co-ordinate zero. That is, the line ℓ_∞ maps to the line $\langle 0, 0, 1 \rangle$ in $\Pi(V)$.

The points on the line $x = b$ along with its point at infinity correspond precisely to the points of $\Pi(V)$ of the form $[b, y, 1]$ and $[0, 1, 0]$ which comprise the line $\langle 1, 0, -b \rangle$ of $\Pi(V)$ (obtained by substituting in points).

The points on the line $y = mx + b$ along with its point at infinity correspond precisely to the points of $\Pi(V)$ of the form $[x, mx + b, 1]$ and $[1, m, 0]$ which comprise the line $\langle -m, 1, -b \rangle$ in $\Pi(V)$. □

Since $\Pi(V)$ is Desarguesian (theorem 3.7.4), it follows that $A(F)$ is Desarguesian.

Finally, let us look at the counterpart of the Pappus theorem in $A(F)$.

Theorem 4.6.4. *Let x, y, z and x', y', z' be sets of three distinct collinear points on distinct lines such that no one of these points is on both lines. Then F is a field if and only if the following always hold: $xy' \| x'y$ and $xz' \| x'z$ imply $yz' \| y'z$; if all points $xy' \cap x'y$, $xz' \cap x'z$, $yz' \cap y'z$ exist, then they are collinear; if $u = xy' \cap x'y$ and $v = xz' \cap x'z$ exist but $yz' \| y'z$, then $uv \| yz'$.*

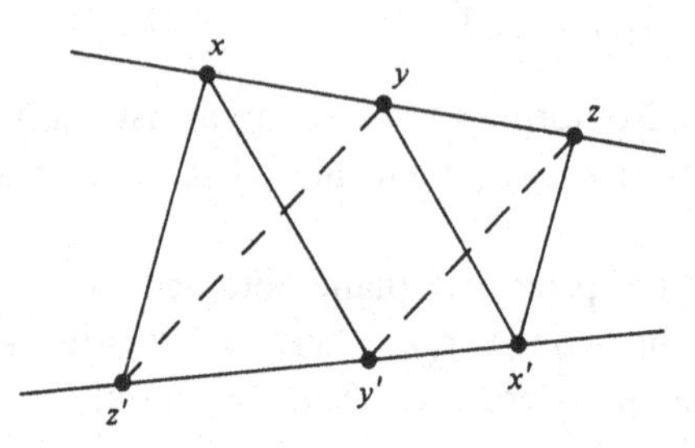

Figure 4.6.2.

Proof. Embed the affine plane $A(F)$ in the projective plane $\Pi(A(F))$ as described above. Then the three points $u = xy' \cap x'y$, $v = xz' \cap x'z$ and $w = yz' \cap y'z$ in $\Pi(A(F))$ are on ℓ_∞ and are therefore collinear. They are collinear if and only if F is a field, by theorem 3.8.1, which finishes the proof. □

The configuration described in theorem 4.6.4 is called an *affine Pappus* configuration, and an affine plane in which the equivalent conditions of the theorem always hold is said to be *Pappian*. Because of theorem 3.8.2, if A is Pappian, it is Desarguesian.

4.7 Affine spaces

Recall (lemmas 4.1.3 and 4.2.1) that all affine planes of order other than 2 actually are planes, that is, have dimension 2. We want to generalize the notion of two-dimensional affine space (affine planes) to n-dimensional affine space. In order to do this we have to have some prior ideas about the kinds of properties we want affine spaces to have. We defined projective spaces (section 3.9) as linear spaces in which each plane is a projective plane. The corresponding definition for affine space would then be a linear space in which each plane is an affine plane. Remember also that an affine plane can always be obtained from a projective plane by removing a line (theorem 4.3.2). Thus, it would be nice if an affine space of dimension n could be obtained from a projective space of dimension n by removing a hyperplane.

So far we have two possible definitions. There are others that attempt to define affine space by means of conditions on 'parallelism' for instance. (See Lenz (1954) and Sasaki (1952) for examples.)

It turns out that the potential definitions we gave above are not the same. Since the second definition, in terms of projective space, makes it much easier for us to collect information about affine space, that is the one we shall use.

An *affine space* is a projective space less a hyperplane.

Example 4.7.1. Consider the three-dimensional projective space of order 2. It has 2^3+2^2+2+1 points. If we remove a hyperplane (a plane), we lose 2^2+2+1 points, as well as all the lines of the plane. The affine space obtained has 2^3 points.

Example 4.7.1 gives us the key to counting the number of points in an affine space. Note first that we have the next result.

Lemma 4.7.1. *All lines in an affine space have the same number of points.*

Proof. By lemma 3.9.6, every line in a projective space meets every hyperplane. Thus every line not in a fixed hyperplane loses exactly one point when the hyperplane is removed. □

If each line in an affine space A has k points, we say A has *order* k.

Lemma 4.7.2. *If the affine space A is obtained from an n-dimensional projective space of order k, then A has k^n points.*

Proof. By lemma 3.9.4, A has $k^n+k^{n-1}+\cdots+k+1-(k^{n-1}+k^{n-2}+\cdots+k+1)=k^n$ points. □

We leave the proof of the following lemma as an exercise.

Lemma 4.7.3. *If the affine space A is obtained from an n-dimensional projective space S, and if the order of S is not* 2, *then A has dimension n.*

Paralleling the notation $P(n,k)$ for projective n-space of order k obtained from a field GF(k), we use $A(n,k)$ for an affine n-space of order k obtained from $P(n,k)$ by deleting a hyperplane.

A projective 3-space of order k has $k^4+k^3+2k^2+k+1$ lines. (See exercise 38 of section 3.11.) So affine 3-space has $k^4+k^3+2k^2+k+1-(k^2+k+1)=k^4+k^3+k^2$ lines.

What does the term 'parallel' mean in affine space? We investigate this a little now. Let p be a point of the hyperplane H removed from the projective space S to get the affine space A. Any two lines of A on p are coplanar in S since they meet in S. We shall say that ℓ and ℓ', lines of A, are *parallel* if they are equal, or if they do not meet in A but do meet in S. The reader can check that, if A is an affine plane, then the old notion of parallelism is the same as the one we have just introduced here. We use the same notation $\parallel$.

Lemma 4.7.4. *Parallelism in an affine space is an equivalence relation.*

Proof. The reflexive property is trivial, and so is the symmetric property.

To check transitivity, let $\ell \parallel \ell'$ and $\ell' \parallel \ell''$. So $\ell \cap \ell' = p \in H$ and $\ell' \cap \ell'' = q \in H$ where H is the hyperplane such that $A = S \setminus H$ for some projective space S. If $p \neq q$, then $\ell' = pq \subseteq H$ which is a contradiction. □

Lemma 4.7.5. *An affine plane (two-dimensional subspace) of an affine space $A = S \setminus H$ is precisely a plane Π of the projective space S, $\Pi \nsubseteq H$, less a line of the hyperplane H of S.*

Proof. Consider a projective plane Π of S as above. By exercise 52 of section 3.11, $\Pi \cap H = \ell$, a line. By theorem 4.3.2, $\Pi \setminus \ell$ is an affine plane.

Conversely, let X be an affine plane of A. Take two parallel classes $[\ell]$ and $[\ell']$ of X. These correspond to distinct points of H, say p and q. Let $\hbar$ be any line of X not in $[\ell]$ or $[\ell']$. Then $\hbar$ meets ℓ and ℓ', and we may assume that it does so in distinct points, x and x' respectively. Let $\ell \cap \ell' = y$, so that x, y and y' are distinct, non-collinear points. Then x, x' and y generate in S a projective plane Π containing ℓ and ℓ', and so containing X. The point $[\hbar]$ in S is therefore on the line pq, and $\Pi = X \cup pq$, with $pq \subseteq H$. □

Lemma 4.7.6. *Every two-dimensional space of an affine space A of order not 2 is an affine plane.*

Proof. Take three non-collinear points of A. Let $A = S \setminus H$. The three points generate a projective plane Π in S. Let ℓ and ℓ' be distinct lines forming two sides of the 'triangle' made by the three points. Let $p = [\ell]$ and $q = [\ell']$ be points of Π and let $\ell_\infty = pq$ in Π. We shall show that the three points generate $\Pi \setminus \ell_\infty$. Let x be any point of this set, not on any side of the triangle. As the order of A is greater than 2, there are at least four lines on x in Π. Excluding the line on x and $\ell \cap \ell'$, the line on x and $[\ell]$ and the line on x and $[\ell']$, any other line on x meets ℓ and ℓ' in distinct points. Hence $x \in \langle \ell \cup \ell' \rangle$ in A. So the two-dimensional space is indeed $\Pi \setminus \ell_\infty$.

The desired conclusion now follows from lemma 4.7.5. □

Corollary. *Two distinct lines of A are parallel if and only if they are in a common affine plane of A and do not meet.*

Lemma 4.7.7. *Each plane of an affine n-space, $n \geq 3$, is Desarguesian.*

Proof. This follows from lemma 4.7.5 and the corollary to lemma 3.10.1. □

Returning to our comments at the beginning of this section, we ask about a 'converse' to lemma 4.7.6. That is, if each plane of a linear space is affine, is the space an affine space? For a counter-example, we refer the reader to M. Hall (1960).

We do have a converse, however, with a simple additional condition. Buekenhout (1969*a*) has proved the following theorem which we state without proof.

Theorem 4.7.8. *Let L be a linear space such that*

(i) *every plane of L is an affine plane,*
(ii) *L has at least three non-collinear points,*
(iii) *every line of L has at least four points.*

Then L is an affine space.

Thus, affine spaces of order at least four can be characterized as above.

4.8 Exercises

1. Draw an affine plane with 16 points.
2. Which of the axioms of an affine plane have valid duals in the affine plane?
3. Show that there is no affine plane with 29 points.
4. Show that the affine plane of order 2 is unique.
5. Show that any affine plane satisfies the exchange property.
6. Show that any affine plane satisfies P2.
7. Prove that in any *near-linear* space satisfying A1, parallelism is an equivalence relation.
8. Let S be a linear space satisfying A1. What can S be?
9. Show that an affine plane always has an even number of lines.
10. If S is a linear space with k^2 points, k points per line and at most $k+1$ lines on each point, describe S.
11. If S is a linear space with k points per line and k^2+k lines, is S an affine plane?
12. Embed the affine plane of order 3 in the projective plane of order 3.

13. Prove lemma 4.2.1 using the results of theorems 4.3.1 and 4.3.2.

14. Prove lemma 4.2.3 using the results of theorems 4.3.1 and 4.3.2.

15. Let C_T be the set of collineations of a linear space S such that $g(T) = T$ for every $g \in C_T$, T a subset of the points of S. Let $\bar{C}_T$ be the set of collineations of S such that $g(t) = t$ for each $t \in T$. Show that $\bar{C}_T$ is a subgroup of C_T which is in turn a subgroup of the group $C(S)$ of all collineations of S.

16. Let the set T of exercise 15 be the set of points of a fixed line l of a projective plane Π. What is the relationship between C_T, $\bar{C}_T$ in Π and the set of collineations of $\Pi \backslash l$?

17. A *normal subgroup* N of a group G is a subgroup of G with the property that $gng^{-1} \in N$ for each $g \in G$ and $n \in N$. We write $N \lhd G$. Using the notation of exercise 15, is $\bar{C}_T \lhd C_T$, $\bar{C}_T \lhd C(S)$ or $C_T \lhd C(S)$?

18. Give an example of a translation of the affine plane $\mathbb{R}^2$ different from that of example 4.4.3.

19. Define $f((x, y)) = (ax, by)$, $ab \neq 0, a$ and b real constants, for each point (x,y) of $\mathbb{R}^2$. Embed $\mathbb{R}^2$ in a projective plane Π and explain how f can be extended to Π.

20. If $D(A)$ is the set of dilatations of the affine plane A and $C(A)$ is the set of all collineations of A, show that $D(A) \lhd C(A)$.

21. Find non-identity collineations of the affine planes with 3^2 and 4^2 points.

22. Let f be a collineation of the affine plane A. Embed A in the projective plane Π and show how to extend f to a collineation of Π.

23. Find a non-identity homothety of the affine plane of order 3.

24. Show that the affine plane of order 2 has no non-identity homothety.

25. Decide which of the following are groups:
(a) the set of homotheties of an affine plane,
(b) the set of translations of an affine plane.

26. If f_Π is a central collineation of Π with axis ℓ, check that f, the function f_Π restricted to $A = \Pi \backslash \ell$ is a dilatation of A.

27. Find an example of a non-identity affine perspectivity in the affine plane of order 4.

28. Let Π be a projective plane and let f be a central collineation of Π. Remove a line ℓ in such a way that the function f restricted to $A = \Pi \backslash \ell$ is no longer a function in A.

29. Show that a translation is uniquely determined by its effect on a single point.

30. Show that a homothety is uniquely determined by its centre and its effect on one other point.

31. Show that an affine perspectivity is uniquely determined by its axis and its effect on a single point not on the axis.

32. Complete the details for the second half of the proof of lemma 4.5.3.
33. Complete the proof of part (viii) of lemma 4.5.2.
34. Prove lemma 4.6.1.
35. Construct $A(F)$ where $F = \mathrm{GF}(k)$ for $k = 2$ and $k = 5$.
36. Find a Desargues configuration in the affine plane $A(2, 3)$.
37. We construct a space $S = (P, L)$ as follows. Let P be the set of points (x, y) of Euclidean 2-space $\mathbb{R}^2$. Any line of $\mathbb{R}$ with non-negative or infinite slope is a line of L. The other lines are $\{(x, y) \mid y = mx + b$ for $x \geq 0$, $y = \frac{1}{2}mx + b, x < 0$ where $m < 0\}$. (Draw some pictures!) Prove that S is a non-Desarguesian affine plane, and hence that the projective plane in which it is embeddable is not Desarguesian (Moufang, 1931).
38. In the affine plane $A(2, 5)$, find a dilatation with centre $(1, 1)$ mapping $(0, 0)$ to $(2, 2)$.
39. An *affinity* of the plane $A(F)$, F a field, is a function of the form f: $(x, y) \rightarrow (ax + by + c,\ a'x + b'y + c')$ where $a, b, c, a', b', c' \in F$ and $ab' - a'b \neq 0$. Show that an affinity is a collineation.
40. Show that the set of affinities of the plane $A(F)$, F a field, is a subgroup of the set of all collineations of $A(F)$.
41. Show that if f is an affinity of $A(F)$ with $c = c' = 0$ and f fixes x and y, $x \neq y$, then $f(z) = z$ for every point z of the line xy.
42. Let $\mathbb{C}$ be the field of complex numbers. That is $\mathbb{C} = \{a + b\mathrm{i} \mid a, b \in R$, the set of real numbers, and $\mathrm{i} = \sqrt{-1}\}$. Define for any $z \in \mathbb{C}$, $z = a + b\mathrm{i}$, the *complex conjugate* of z: $\bar{z} = a - b\mathrm{i}$. In $A(\mathbb{C})$ show that the function f: $(z, z') \rightarrow (\bar{z}, \bar{z}')$ is a collineation of $A(\mathbb{C})$. (Hint: show $\overline{xy + zw} = \bar{x}\bar{y} + \bar{z}\bar{w}$.)
43. Find a Pappus configuration (figure 4.6.2) in the affine plane $A(2, 3)$.
44. If the projective space S has dimension 1, what is the corresponding affine space?
45. Prove lemma 4.7.3.
46. Find the number of lines and the number of hyperplanes in affine n-space.
47. Find the number of planes on a given line in affine n-space.
48. Find the number of planes in affine 4-space.
*49. Find a new definition of affine space. (It should be equivalent to our old definition.)
50. Show that any two planes in an affine 3-space either meet in a line or don't meet at all. (See axiom A4 of Sasaki (1952).)
51. What can you say about the intersections of hyperplanes in an affine d-space?
*52. Develop a theory of affine subplane similar to that of projective subplane in chapter 3.

53. A *Latin square* of order n, $n \geq 1$, is an $n \times n$ array of integers such that each row and each column of the array contains each integer between 1 and n (inclusive) exactly once. The following matrix is an example with $n = 3$.

$$\begin{bmatrix} 1 & 2 & 3 \\ 2 & 3 & 1 \\ 3 & 1 & 2 \end{bmatrix}$$

The Latin squares

$$A = \begin{bmatrix} a_{11} & \cdots & a_{1n} \\ \vdots & & \vdots \\ a_{n1} & \cdots & a_{nn} \end{bmatrix} \text{ and } B = \begin{bmatrix} b_{11} & \cdots & b_{1n} \\ \vdots & & \vdots \\ b_{n1} & \cdots & b_{nn} \end{bmatrix} \text{ of order } n \text{ are}$$

orthogonal if all n^2 different ordered pairs (a, b), $1 \leq a, b \leq n$, appear as ordered pairs (a_{ij}, b_{ij}), $1 \leq i, j \leq n$. It is known (Hall (1967)) that there is an affine plane of order n if and only if there is a set of $n - 1$ pairwise orthogonal Latin squares of order n. Find such a set of Latin squares for $n = 5$.

***54.** Try exercise 53 for $n = 10$. (See section 3.2.)

5

Polar spaces

How will you know the pitch of that great bell
Too large for you to stir? Let but a flute
Play 'neath the fine-mixed metal listen close
Till the right note flows forth, a silvery rill
Then shall the huge bell tremble – then the mass
With myriad waves concurrent shall respond
In low soft unison.

George Eliot *Middlemarch*

Polar spaces as represented here are a very recent invention. The definition we give is due to Buekenhout and Shult (1974). They were first defined by Veldkamp (1959) and the theory has since then been greatly developed by Tits (1974). Also, much of the material here can be found in Buekenhout and Shult (1974) and Buekenhout and Deherder (1971). We outline the history of the topic in section 5.7.

5.1 The definition

A *polar space* $S = (P, L)$ is a near-linear space such that for every point p not in the line ℓ, $c(p, \ell) = 1$ or $v(\ell)$.

Example 5.1.1. Any linear space is a polar space and, in particular, affine and projective spaces are polar spaces.

Example 5.1.2. The near-linear spaces of figure 5.1.1 are examples of polar spaces.

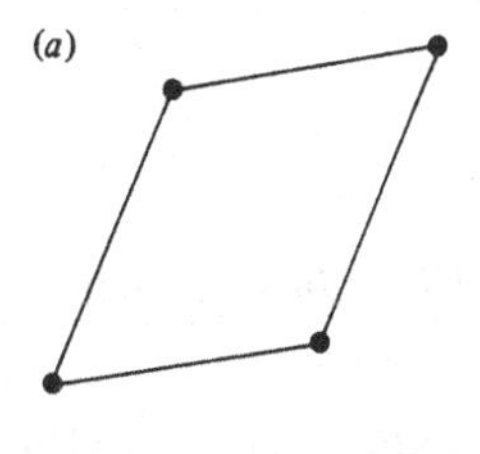

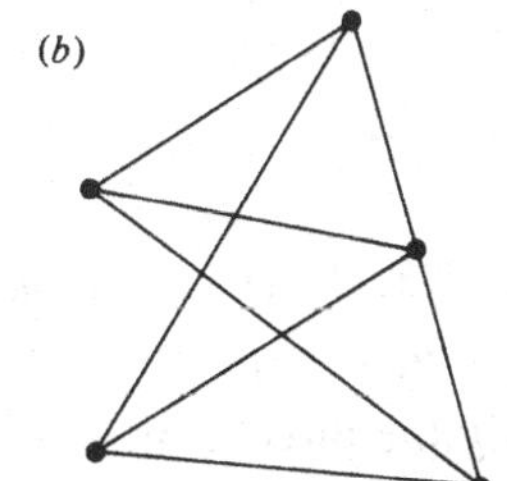

Figure 5.1.1.

Sections 5.2 and 5.3 will be devoted to developing theory that shows the existence of some very interesting classes of examples of polar spaces.

If some point of S is collinear with all points of S, we shall say that S is *degenerate*. We shall also say that $\varnothing$ is degenerate. In particular, all linear spaces are degenerate polar spaces. The example of figure 5.1.1 (*b*) is degenerate while that of figure 5.1.1 (a) is non-degenerate.

We shall see that non-degenerate polar spaces have particularly nice properties that are often lost in the degenerate case.

It is a simple matter to produce new polar spaces from old ones by using direct sums:

Let $\{S_i\}$, i in some index set I, be a family of pairwise disjoint polar spaces. The *direct sum* of the S_i is defined to be the space S whose points are the points of the S_is and whose lines are the lines of the S_is along with all pairs of points $\{p, q\}, p \in S_i, q \in S_j, i \neq j$. We use the notation $S = \bigoplus S_i$ where i is assumed to run through an appropriate index set. Essentially we are 'sticking' polar spaces together by joining each point of one to every point of the others.

The space of figure 5.1.2 (*c*) is the direct sum of the polar spaces of figures 5.1.2 (*a*) and (*b*) where the dotted lines represent the new 2-point lines.

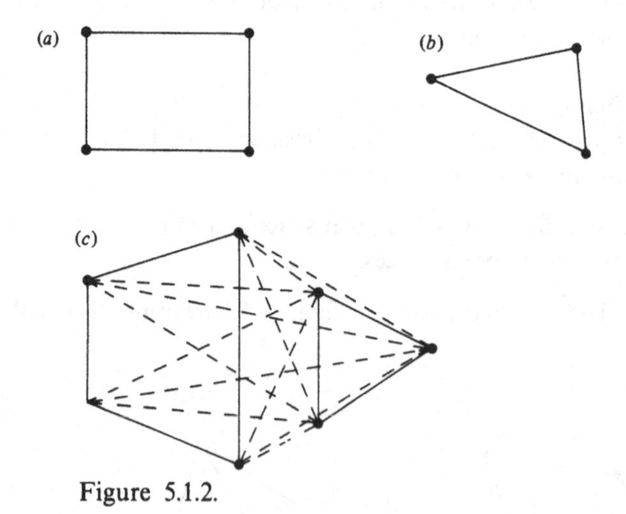

Figure 5.1.2.

Lemma 5.1.1. *The direct sum of any family $\{S_i\}$, i in some index set I, of polar spaces, is again a polar space.*

Proof. Clearly the direct sum S is still a near-linear space. Let p be any point and ℓ any line of S, $p \notin \ell$. If p and ℓ are in the same S_i, then $c(p, \ell) = 1$ or $v(\ell)$ by definition. If $p \in S_i$ and $\ell \in S_j$, $i \notin j$, then $c(p, \ell) = v(\ell)$ by definition of S. If $p \in S_i$ and ℓ is a new 2-point line, then p is joined to at least one point of ℓ. □

It is also possible to get new polar spaces from old ones simply by taking subspaces. We leave it as an exercise to show that any subspace of a polar space is again a polar space.

Lemma 5.1.2. *Let $S = \bigoplus S_i$ be a direct sum of polar spaces. Then any subspace R of S is a direct sum of polar spaces $R = \bigoplus R_i$ where $R_i = R \cap S_i$.*

Proof. Clearly each element of R_i is adjacent to all elements of R_j, $i \neq j$. Furthermore, $R_i = R \cap S_i$ as an intersection of subspaces is a subspace. And any subspace of a polar space is a polar space. □

Lemma 5.1.3. *If M is a maximal subspace of the polar space $S = \bigoplus S_i$, then $M_i = M \cap S_i$ is maximal in S_i for all i.*

Proof. Suppose M_j is not maximal in S_j for some j. Let $M_j \subsetneq \bar{M}_j \subsetneq S_j$. It is easily seen then that $M = \bigoplus M_i$ (by lemma 5.1.2) $\subsetneq (\bigoplus_{i \neq j} M_i) \oplus \bar{M}_j \subsetneq \bigoplus S_i = S$ contradicting M maximal in S. □

5.2 Absolute points

The theory of polar spaces arose in connection with the set of absolute points of non-degenerate polarities in Desarguesian projective spaces and with non-degenerate quadrics in projective spaces over fields. In this section and the next we shall consider these two aspects. Most of the results of this section will concern polarities in projective planes, while the final part of the section will include higher-dimensional projective spaces.

A *correlation* σ on a d-dimensional projective space S, d finite, is a 1–1 map of the set of subspaces of S into itself such that, for all subspaces R and T, $R \subseteq T$ implies $\sigma(R) \supseteq \sigma(T)$ and $\dim \sigma(R) = d - 1 - \dim R$.

Thus a correlation reverses inclusion and R is the 'same distance' from $\varnothing$ as $\sigma(R)$ is from S. (Compare this definition with the definition we gave for projective *planes* in exercise 19 of section 3.11.)

A correlation of *order* 2, that is, satisfying $\sigma^2 = \sigma \circ \sigma = 1$, the identity map, is called a *polarity*.†

Note that, if σ is a polarity, then $R \subseteq \sigma(T)$ implies $T \subseteq \sigma(R)$. In particular, for points p and q, $p \in \sigma(q)$ implies $q \in \sigma(p)$, and $\sigma(p)$ and $\sigma(q)$ are hyperplanes of S.

An *absolute subspace* of a polarity σ is a subspace R satisfying $R \supseteq \sigma(R)$ or $R \subseteq \sigma(R)$. Clearly, if R is absolute then so is $\sigma(R)$ and conversely.

† It has been shown (Tits, 1974) that the existence of a polarity on a projective space S implies $\dim S$ finite.

Example 5.2.1. Consider the Fano plane labelled as in figure 1.1.1. Define σ:

$0 \to 045$
$1 \to 235$
$2 \to 124$
$3 \to 156$
$4 \to 026$
$5 \to 013$
$6 \to 346$

If σ has order 2, each line must map back to the point from which it came. This is a polarity and the reader is left to check the details.

It is easy to see that the absolute points are 0, 2 and 6 and so the absolute lines are 045, 124 and 346.

The results of lemmas 5.2.1 and 5.2.2 and of theorem 5.2.3 apply only to finite projective planes. Hence the hyperplanes are the lines.

Lemma 5.2.1. *Every absolute line of the polarity σ of a projective plane of finite order contains precisely one absolute point; dually, every absolute point is on precisely one absolute line.*

Proof. Suppose ℓ is absolute. Thus the point $\sigma(\ell)$ on ℓ is absolute also. If ℓ contains two (or more) absolute points, say p and q, then $p, q \in \ell$ implies $\sigma(\ell) \in \sigma(p)$, $\sigma(q)$. Also, $p \in \sigma(p)$ and $q \in \sigma(q)$. So $\sigma(p)$ contains $\sigma(\ell)$ and p, both on ℓ, and $\sigma(q)$ contains $\sigma(\ell)$ and q, both on ℓ. Since the point $\sigma(\ell)$ cannot be *both* p and q, we get $\sigma(p) = \sigma(q) = \ell$ while σ 1–1 implies $p = q$, a contradiction. □

Corollary. *A polarity of a finite projective plane always has non-absolute points.*

Lemma 5.2.2. *Every non-absolute line in a projective plane of finite order contains an even number of non-absolute points.*

Proof. Let ℓ be a non-absolute line, and so $\sigma(\ell) \notin \ell$. Let p be any point of ℓ. Then p is absolute if and only if $p \in \sigma(p)$. But since σ is a correlation, $p \in \ell$ implies $\sigma(\ell) \in \sigma(p)$. Hence, $p \in \sigma(p)$ if and only if $p = \sigma(p) \cap \ell$.

Thus the set of non-absolute points of ℓ falls into pairs (see figure 5.2.1): p is associated with $p' = \ell \cap \sigma(p)$, so there is an even number of non-absolute points. □

Corollary 1. *If the order k of the projective plane is even, then every line contains an odd number of absolute points.*

Proof. If the line ℓ is absolute, it has precisely one absolute point by lemma

5.2.1. If ℓ is non-absolute, it follows from lemma 5.2.2 that its number of absolute points is odd. □

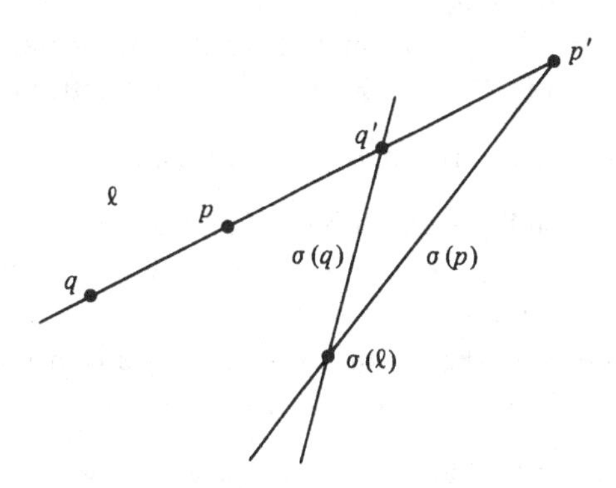

Figure 5.2.1.

Corollary 2. *If the order k of the projective plane is odd, then a line ℓ is absolute if and only if it contains exactly one absolute point.*

Proof. Lemma 5.2.1 tells us that an absolute line has exactly one absolute point. In the other direction, if ℓ is non-absolute and has precisely one absolute point, we contradict lemma 5.2.2. □

The next theorem is due to Baer (1946*a*) and it gives us some information about the number of absolute points possible for a given polarity. We state it without a proof.

Theorem 5.2.3. *Let t be the total number of absolute points of the polarity σ on the projective plane Π of order k. Then $t \equiv k+1 \pmod 2$, and if p is any odd prime and i any non-negative integer we have*

$$(t-k-1)k^{(p^i-1)/2}(k^{(p-1)p^i/2}-1) \equiv 0 \bmod (p^{i+1}).$$

The second congruence places a large restriction on the number of values possible for t with respect to a given k.

Consider again the general case of a d-dimensional projective space S, d finite. A subspace R of S will be called *totally isotropic, isotropic* or *non-isotropic* with respect to the polarity σ as $R \cap \sigma(R) = R$, $\neq \emptyset$, $= \emptyset$, respectively. Note that if $d = 3$, then the totally isotropic lines are those lines ℓ for which $\sigma(\ell) = \ell$.

We are interested especially in totally isotropic lines, and prove the following result.

Lemma 5.2.4. *In a finite-dimensional projective space with polarity σ, a line is totally isotropic if and only if for any p on ℓ, $p \in \sigma(q)$ for all q on ℓ.*

Proof. Suppose the line ℓ is totally isotropic. Let q be any point of ℓ. (This includes the case $q=p$.) Then $q \in \ell$ implies $\sigma(\ell) \subseteq \sigma(q)$, while ℓ totally isotropic gives $\ell \subseteq \sigma(\ell)$. So $p \in \ell \subseteq \sigma(q)$.

Now suppose $p \in \sigma(q)$ for all p and q on ℓ. Then $\ell \subseteq \sigma(q)$ for all $q \in \ell$. Thus $\sigma(\ell) \ni q$ for all $q \in \ell$. That is, $\ell \subseteq \sigma(\ell)$. Hence ℓ is totally isotropic. □

Corollary. *Each point of a totally isotropic line is absolute.*

Notice that in the case $\dim S=2$ and S has finite order, lemmas 5.2.1 and 5.2.4 imply that there are no totally isotropic lines. In the case of dimension greater than 2, such lines may exist.

We now restrict ourselves to the set of absolute points of a polarity and, equipped with appropriate lines, we have a polar space:

Theorem 5.2.5. *Let S be a finite-dimensional projective space with polarity σ. Then the set of absolute points of σ equipped with the totally isotropic lines of σ form a polar space.*

Proof. Let ℓ be a totally isotropic line and p an absolute point not on ℓ. Let x be any point on ℓ. Then the line xp is totally isotropic if and only if $x \in \sigma(p)$ by lemma 5.2.4. Now $\sigma(p)$ is a hyperplane and by lemma 3.9.6 must meet ℓ in a single point or in ℓ itself. Thus either p is adjacent to one point of ℓ, or to all of them. □

The polarity σ of example 5.2.1 yields a polar space with three points and no lines.

Example 5.2.2. Let S be three-dimensional projective space over the complex numbers F. Define

$$\sigma: p=[p_0,p_1,p_2,p_3] \leftrightarrow \sum_{i=0}^{3} p_i x_i = 0.$$

Then p is in $\sigma(p)$ precisely when $\sum_{i=0}^{3} p_i^2 = 0$.

If $p \in \sigma(p)$ and $q \in \sigma(q)$, $p \neq q$, then any point $a=[a_0,a_1,a_2,a_3]$ of pq has the form $a=\lambda p+\mu q$, $\lambda, \mu \in F$. But

$$\sum_{i=0}^{3} a_i x_i = \lambda \sum_{i=0}^{3} p_i x_i + \mu \sum_{i=0}^{3} q_i x_i = 0$$

implies $a \in \sigma(a)$ so that every point of such a line is absolute. All such lines then are totally isotropic.

5.3 Quadrics

In section 3.7, where we introduced fields, we mentioned that the integers modulo a prime were always a field. We also said that it was possible to construct fields of prime power order, say p^n, where the operations of addition, etc., were still performed modulo p. In fact, the operations of *any* finite field can be shown to be performed essentially modulo some prime p. (See Curtis (1967) for instance.) This number p is called the *characteristic* of the field.

A *quadric* Q in a projective space S of dimension $d \geq 0$ over the field F is the set of points of S satisfying a second degree homogeneous (that is, the sum of the degrees of each term is a constant) polynomial equation in $d + 1$ variables over F:

$$a_{00}x_0^2 + a_{01}x_0x_1 + a_{02}x_0x_2 + \cdots + a_{dd-1}x_dx_{d-1} + a_{dd}x_d^2 = 0,$$

$a_{ij} \in F$, or more briefly

$$\sum_{i=0}^{d} \sum_{j=0}^{d} a_{ij}x_ix_j = 0.$$

It will be convenient to use matrix notation to represent the above polynomial equation. Notice that, since $x_ix_j = x_jx_i$, there are several ways of interpreting the coefficients $a_{ij} + a_{ji}$. For instance, $x_1^2 + x_0x_2 = 0$ could be represented in matrix form by

$$\begin{pmatrix} 0 & 0 & 1 \\ 0 & 1 & 0 \\ 0 & 0 & 0 \end{pmatrix} \quad \text{or by} \quad \begin{pmatrix} 0 & 0 & \frac{1}{2} \\ 0 & 1 & 0 \\ \frac{1}{2} & 0 & 0 \end{pmatrix}$$

as a quadric over the reals with $d = 2$.

If the characteristic of the field were not 2, we could always require the unique *symmetric* matrix, which in the above would be the latter. However, this is not essential for what we want to do here.

In any case, if $x = [x_0, x_1, \ldots, x_d]$ is a $1 \times (d + 1)$ matrix (a vector),

$$x^t = \begin{bmatrix} x_0 \\ x_1 \\ \cdot \\ \cdot \\ \cdot \\ x_d \end{bmatrix}$$

(a $(d + 1) \times 1$ matrix) is the transpose of x and A is any matrix representing the coefficients of the equation of the quadric, then that equation can be written as $xAx^t = 0$.

We now examine the possibilities for the intersection of a line with a quadric.

Let Q be a quadric determined by the equation

$$\sum_{i=0}^{d} \sum_{j=0}^{d} a_{ij}x_ix_j = 0.$$

Let ℓ be a line determined by points $p=[p_0,p_1,\ldots,p_d]$ and $q=[q_0,q_1,\ldots,q_d]$. The line ℓ is therefore the set of points which are linear combinations of p and q. Thus, for any point x of ℓ, $x=[x_0,x_1,\ldots,x_d]$, we have equations $x_i=\lambda p_i+\mu q_i$ for λ, $\mu\in F$ not both zero.

The intersection of Q and ℓ is therefore the set of these points x which are also on Q. Thus the following must hold:

$$\sum_{i=0}^{d}\sum_{j=0}^{d} a_{ij}(\lambda p_i+\mu q_i)(\lambda p_j+\mu q_j)=0$$

or equivalently

$$\lambda^2\left(\sum_i\sum_j a_{ij}p_ip_j\right)+\lambda\mu\left(\sum_i\sum_j(a_{ij}+a_{ji})p_iq_j\right) + \mu^2\left(\sum_i\sum_j a_{ij}q_iq_j\right)=0 \qquad (*)$$

Example 5.3.1. Consider the quadric Q:

$$x_0^2+2x_1^2+3x_2^2=0$$

in two-dimensional space over the field $F=\mathrm{GF}(5)$, and the line ℓ on $p=[1,0,1]$ and $q=[1,2,0]$. Any point x in the intersection of Q and ℓ satisfies

$$x=\lambda[1,0,1]+\mu[1,2,0], \text{ and also}$$
$$4\lambda^2+2\lambda\mu+4\mu^2=0.$$

If $\mu=0$, then $\lambda=0$. So we may rewrite this as

$$4(\lambda\mu^{-1})^2+2(\lambda\mu^{-1})+4=0 \quad \text{or}$$
$$2(\lambda\mu^{-1})^2+\lambda\mu^{-1}+2=0.$$

The only solution in GF(5) is $\lambda=\mu$ and so $x=[2,2,1]$.

We are able in fact to determine the number of points in the intersection of a line and quadric.

Lemma 5.3.1. *Let Q be a quadric in projective d-space over the field F and ℓ a line. Then $|\ell\cap Q|=0$, 1, 2 or $v(\ell)$.*

Proof. We show that if $|\ell\cap Q|>2$, then $\ell\subseteq Q$. Let p, $q\in\ell\cap Q$. We may suppose these are the points p and q of (*). Thus the terms with λ^2 and μ^2 are zero. If the coefficient of $\lambda\mu$ is also zero, then all choices of λ and μ satisfy, so that all points of ℓ are in Q. If the coefficient of $\lambda\mu$ is not zero, then either λ or $\mu=0$ and the non-zero variable may be assumed to be 1. Thus p and q are the only solutions here, a contradiction. □

If Q is a quadric in projective d-space over F and if p is a point of Q, we define a *tangent* to Q at p as a line which meets Q just at p, or which is completely contained in Q. This digresses somewhat from the usual definition of tangent to a curve in Euclidean space, say. But the fact that lines can be completely contained within quadrics in our situation slightly changes our outlook. We shall see that it is more convenient to use the definition we have just given.

In Euclidean 3-space, a tangent plane to a curve at a point is just the union of the tangent lines to the curve at that point. This fact motivates us to consider the *union Q_p of all tangents to Q through p.*

Lemma 5.3.2. *If Q is a quadric in projective d-space over the field F, with matrix $A=(a_{ij})$, and if $p=[p_0,p_1,\ldots,p_d]\in Q$, then Q_p is a hyperplane or the whole space and has equation*

$$\sum_{j=0}^{d}\sum_{i=0}^{d}((a_{ij}+a_{ji})p_i)x_j=0.$$

Proof. Let $q\neq p$ be any point of Q_p. So the line qp is tangent to Q at p. Using (*), $p\in Q$ implies $\sum_i\sum_j a_{ij}p_ip_j=0$. As $\mu=0$ corresponds to the solution p, we may suppose $\mu\neq 0$ in which case (*) reduces to

$$\lambda\left(\sum_i\sum_j(a_{ij}+a_{ji})p_iq_j\right)+\mu\left(\sum_i\sum_j a_{ij}q_iq_j\right)=0$$

or

$$\lambda\mu^{-1}\left(\sum_i\sum_j(a_{ij}+a_{ji})p_iq_j\right)=-\sum_i\sum_j a_{ij}q_iq_j,$$

which is a linear equation in $\lambda\mu^{-1}$. It follows that either all elements of F satisfy (so $pq\cap Q=pq$) or precisely one element of F satisfies (so $pq\cap Q=\{p\}$). Hence, the line pq is tangent to Q if and only if

$$\sum_i\sum_j(a_{ij}+a_{ji})p_iq_j=0.$$

This equation, however, is a homogeneous linear equation in the q_j and therefore represents either a hyperplane of the space or the space itself. □

Lemma 5.3.2 enables us to give very quickly another class of polar spaces derived from projective spaces.

Theorem 5.3.3. *Let Q be a quadric in projective d-space over the field F. Then the points of Q equipped with the lines it contains form a polar space.*

Proof. Let ℓ be a line of Q and p a point of Q not on ℓ. By lemma 5.3.2, Q_p is either a hyperplane or the whole space. These possibilities respectively yield p is adjacent to 1 or to all points of ℓ. □

We shall say that a quadric is *non-degenerate* if as a polar space it is non-degenerate.

Points p and q of Q are said to be *conjugate* with respect to Q with matrix A if $pAq^t = 0$. If A is symmetric, this is equivalent to $qAp^t = 0$. Notice that all points of Q are conjugate to themselves.

Lemma 5.3.4. *Let p and q be distinct points of the quadric Q over a field of characteristic not 2. Then p and q are conjugate if and only if the line pq is contained in Q.*

Proof. Let r be any point of the line pq. As in section 3.7, we may choose field co-ordinates so that $r = p + q$. Then $rAr^t = (p+q)A(p+q)^t = pAp^t + pAq^t + qAp^t + qAq^t$. Since p and q are points of the conic, $pAp^t = qAq^t = 0$. Furthermore, qAp^t is a 1×1 matrix (a field element) and so $qAp^t = (qAp^t)^t = pAq^t$. Therefore, $rAr^t = 2pAq^t$.

If p and q are conjugate, then $pAq^t = 0$ and so $rAr^t = 0$ implying $r \in Q$. Conversely, if $r \in Q$, $pAq^t = 0$ and so p and q are conjugate. □

Example 5.3.2. Consider the quadric determined by the equation $x_1^2 + x_0x_1 + x_1x_2 + x_2^2 + x_3^2 = 0$ in $P(3,3)$. Its matrix is

$$A = \begin{pmatrix} 0 & 2 & 0 & 0 \\ 2 & 1 & 2 & 0 \\ 0 & 2 & 1 & 0 \\ 0 & 0 & 0 & 1 \end{pmatrix}.$$

(Note that $\frac{1}{2} = 2$ in GF(3).) The points $p = [1,1,0,1]$ and $q = [1,2,1,0]$ are on the quadric, but are not conjugate to each other. We leave the reader to compute all points on the quadric, and to show that no two points of it are conjugate to each other.

Example 5.3.3. Let F be the field of real numbers. Let Q be determined by the matrix

$$A = \begin{pmatrix} 1 & -1 & 0 & -1 \\ -1 & 0 & 0 & 1 \\ 0 & 0 & 3 & 0 \\ -1 & 1 & 0 & 1 \end{pmatrix}.$$

The point $p = [1,\frac{1}{2},0,0]$ is on Q. The reader should check that all points of Q conjugate to p have the form $[2x_0, x_0 - x_2, 0, 2x_2]$ for all choices of x_0 and x_2 not both zero.

We give more information about quadrics here without proofs. The reader is referred to Buekenhout (1978) or Hirschfeld (1979) for details.

A non-degenerate quadric in $2d$-dimensional projective space over GF(k) contains subspaces of the projective space of dimension up to and including $d-1$, but no higher. In this case, we call the quadric a *parabolic quadric*. In odd-dimensional projective space, say of dimension $2d-1$, there may be $(d-1)$-dimensional or $(d-2)$-dimensional subspaces of highest dimension. In the former case the quadric is called *hyperbolic* or *ruled*; in the latter case it is called *elliptic* or *unruled*.

The number of different m-dimensional subspaces of a projective space of finite dimension over $GF(k)$ to be found in a non-degenerate quadric is as follows:

(i) dimension $2d$, $\prod_{i=0}^{m}(k^{2(d-m+i)}-1)/(k^{m+1-i}-1)$ and $m \le d-1$;

(ii) dimension $2d-1$, elliptic quadric, $m \le d-2$,

$$\prod_{i=0}^{m} \frac{k^{2d-1-2m+2i}+k^{d-m+i-1}-k^{d-m+i}-1}{k^{m+1-i}-1};$$

(iii) dimension $2d-1$, hyperbolic quadric, $m \le d-1$,

$$\prod_{i=0}^{m} \frac{k^{2d-1-2m+2i}-k^{d-m+i-1}+k^{d-m+i}-1}{k^{m+1-i}-1}.$$

It is also possible to determine the total number of non-degenerate quadrics in such a projective space. This number is

$$k^{d(d+1)}\prod_{i=1}^{d}(k^{2i+1}-1) \quad \text{for dimension } 2d,$$

$$k^{d(d+1)}\prod_{i=1}^{d-1}(k^{2i+1}-1) \quad \text{for dimension } 2d-1.$$

In the second case there are

$$\frac{1}{2}k^{d^2}(k^d+1)\prod_{i=1}^{d-1}(k^{2i+1}-1)$$

ruled quadrics and so

$$\frac{1}{2}k^{d^2}(k^d-1)\prod_{i=1}^{d-1}(k^{2i+1}-1)$$

unruled quadrics.

5.4 Linear subspaces

We wish to look more closely now at special types of subsets occurring in the polar space S. These are *linear subspaces,* or subspaces of S which are linear spaces. Any point or line of S is a linear subspace of S, and of course if S is itself a linear space then every subspace is linear.

The next definition is most common in graph theory, but using it here will allow us to abbreviate statements a good deal.

A subset C of the points of a near-linear space is called a *clique* if any pair of points of C is collinear.

Lemma 5.4.1. *Let S be a polar space and C a maximal clique. Then C is a linear subspace of S.*

Proof. Let p and q be distinct points of C. So p and q are collinear. If C contains no point outside of pq, then $C = pq$ as C is maximal. Otherwise, let $r \in C \backslash pq$. Then $c(r, pq) \geq 2$ so that $c(r, pq) = v(pq)$. Hence every point of $C \backslash pq$ is collinear with each point of pq. Clearly this is also true for $r \in pq$. Since C is maximal, $C \supseteq pq$ and so C is a linear subspace. □

It follows from Zorn's lemma[†] that any clique X is contained in a maximal clique and thus, by lemma 5.4.1, in a linear subspace. We use the notation $L(X)$ for the intersection of all linear subspaces on X, or equivalently for the smallest linear space on X, and will often say that X *generates* $L(X)$. We have the following properties the proofs of which we leave as an exercise: $X \subseteq L(X)$, $L(X) = L(L(X))$, $X \subseteq Y$ implies $L(X) \subseteq L(Y)$. (L acts as a *closure operator*. See exercise 53 of section 2.7.)

Recall that a projective hyperplane of a near-linear space S is a proper subspace which meets every line of S. (We defined this for *linear* spaces in section 2.5.)

We introduce the following notation. Let R be a subspace of the near-linear space S with $p \in S \backslash R$. The set of points of R collinear with p will be denoted by R_p.

Lemma 5.4.2. *If X is a clique, R a linear space, $p \notin R$ and $X \subseteq R_p$ then $L(X) \subseteq R_p$.*

Proof. Since R is a linear space, by definition of $L(X)$ we have $L(X) \subseteq R$. Let $q \in L(X) \backslash X$. Then, since $L(X)$ is the *smallest* linear space on X, $q \in xy$ where x and y are distinct points of X. Since p is collinear with x and y, it is collinear with q by definition of polar space. □

Lemma 5.4.3. *Let S be a polar space and R a subspace such that there is a point $p \in S \backslash R$ with $R_p \neq R$. Then R_p is a projective hyperplane of R.*

Proof. Let ℓ be any line of R. As S is a polar space, $c(p, \ell) = 1$ or $v(\ell)$ and in either case the intersection of ℓ with R_p is non-empty. □

Lemma 5.4.4. *Let S be a polar space and R a maximal linear subspace of S. Suppose that there is a point $p \in S \backslash R$ with $R_p \neq R$. Then $L(R_p \cup \{p\})$ is maximal also and is the union of lines joining p to points of R_p.*

Proof. We shall prove first that R_p is a projective hyperplane of $L(R_p \cup \{p\})$; from this it follows immediately that $L(R_p \cup \{p\})$ is the

[†] Let $\mathscr{S}$ denote a non-empty set of subsets of a fixed set S. If for each chain of subsets of $\mathscr{S}$, $C_1 \subseteq C_2 \subseteq \ldots$, the union $\bigcup_i C_i$ is in $\mathscr{S}$, then $\mathscr{S}$ has a maximal element (Birkhoff, 1967).

union of lines joining p to points of R_p. So, let $q \in R \setminus R_p$. Thus q and p are not collinear and so $q \notin L(R_p \cup \{p\})$. We are therefore able to introduce the set $L(R_p \cup \{p\})_q$ of which R_p is clearly a subset since R is linear.

Suppose there is a point $x \in L(R_p \cup \{p\})_q \setminus R_p$. Then $x \in L(R_p \cup \{p\})$ implies x and p collinear and so $x \notin R$. Since $R_p \cup \{q\}$ is a clique, lemmas 5.4.2 and 5.4.3 imply $R = L(R_p \cup \{q\}) \subseteq R_x$, and $R \supseteq R_x$ implies $x \in R$, a contradiction. Then by lemma 5.4.3, $L(R_p \cup \{p\})_q = R_p$ is a projective hyperplane of $L(R_p \cup \{p\})$.

Suppose $L(R_p \cup \{p\})$ is not maximal, but is contained in a proper linear subspace Y of S. Any line ℓ joining p to a point of $Y \setminus L(R_p \cup \{p\})$ does not meet R_p. (See figure 5.4.1.) But, since S is a polar space, $q \in R \setminus R_p$ is collinear with some point r of ℓ, and clearly $r \in Y \setminus R_p$. Now r is collinear with all points of R_p since Y is linear, so $R_r \supseteq R_p \cup \{q\}$ implying $R_r \supseteq L(R_p \cup \{q\}) = R$ by lemma 5.4.2. Moreover, $r \notin R$ since $r \notin R_p$, and R_p a hyperplane of R would imply

$$L(R_p \cup \{q\}) = R = L(R_p \cup \{r\}) \subseteq Y$$

and so $R \subseteq Y$, a contradiction since R is maximal. Thus $R \cup \{r\}$ is a clique properly containing R. It follows that $L(R \cup \{r\}) = S$ is a linear space while this is false since, for instance, $R_p \neq R$. □

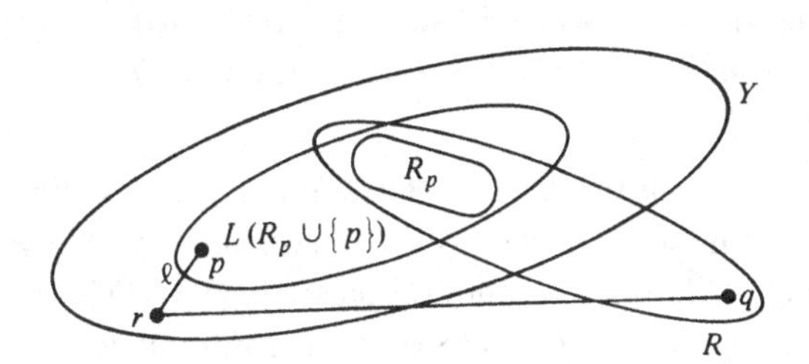

Figure 5.4.1.

We shall say that a polar space S has *finite rank* if every chain

$$\emptyset = L_{-1} \subsetneq L_0 \subsetneq L_1 \subsetneq \cdots, \quad \text{or} \quad L_{n-1} \supsetneq L_{n-2} \supsetneq L_{n-3} \supsetneq \cdots,$$

of distinct linear subspaces has finite length (i.e., a finite number of elements). If the maximum length over all such chains is $n+1$, we say that S has *rank n*.

Lemma 5.4.5. *Let S be a non-degenerate polar space of finite rank. For any linear subspace R of S there is a maximal linear subspace disjoint from R.*

Proof. A maximal linear subspace M of S exists by the finite rank assumption. If $M \cap R = \emptyset$ then we are done, so suppose otherwise. Let $p \in R \cap M$. Since S is non-degenerate, there is a point q that is not collinear with p.

Suppose there is a point $r \in R \setminus M$ such that $r \in L(M_q \cup \{q\})$. Hence $M_q \cup \{p, r\}$ is a clique. By lemmas 5.4.3 and 2.5.6, M_q is a projective hyperplane and therefore a hyperplane of M, and so $L(M_q \cup \{p\}) = M$ and $L(M_q \cup \{p, r\})$ is a linear subspace properly containing M, contradicting M maximal. So no such point r exists.

Now $M_q \neq M$ implies by lemma 5.4.4 that $M' = L(M_q \cup \{q\})$, is a maximal linear subspace of S. Moreover, the above paragraph implies that $M' \cap R \subseteq M \cap R$. In fact, $p \notin M'$ implies $M' \cap R \subsetneq M \cap R$. The lemma is now proved by noting that S has finite rank, and that we can repeat the above argument a finite number of times to get $M^{(s)} \cap R = \emptyset$ for some integer s. □

For any point p, define $p^\perp = \{x | x \text{ and } p \text{ are collinear or } x = p\}$.
The *radical* of S is the set

$$\operatorname{rad} S = \{p | p^\perp = S\}.$$

Thus, by definition, S is non-degenerate if and only if $\operatorname{rad} S = \emptyset$.

Lemma 5.4.6. *Let Q and R be linear subspaces of S, a non-degenerate polar space of finite rank, such that $Q \subseteq R$. Then there is a maximal linear subspace M such that $M \cap R = Q$.*

Proof. We use induction on the rank of Q as a polar space, as this must also be finite. If this number is 0, then $Q = \emptyset$ and lemma 5.4.5 gives the result. So suppose $Q \neq \emptyset$. Since S is non-degenerate, $Q \neq S$. Moreover, if for all $p \in S \setminus Q$, $Q_p = Q$, then $Q \subseteq \operatorname{rad} S = \emptyset$, a contradiction. So, by lemma 5.4.3, there is a projective hyperplane Q' of Q. Let $p \in Q \setminus Q'$. By induction, there is a maximal linear subspace M' of S such that $M' \cap R = Q'$. Note that $M' \cap Q = Q'$ also, and $p \in M'$.

Let $M = L(M'_p \cup \{p\})$, which is a maximal linear subspace of S containing $Q = L(Q' \cup \{p\})$ by lemma 5.4.4. Clearly, $M \cap R \supseteq Q$. M' is maximal while Q' is not, so that we can choose $q \in M' \setminus Q'$. To show $M \cap R = Q$, suppose there is a point $r \in ((M \cap R) \setminus Q)$. As $M \cap R$ is a linear space on both p and r, there is a line pr. M'_p is a projective hyperplane of M by lemma 5.4.3, so that pr meets M'_p in a point $s \neq p$. Moreover, $pr \subseteq R$ implies $s \in R \cap M' = Q'$. Thus pr contains two distinct points of Q, implying $pr \subseteq Q$ while $r \notin Q$. This gives us a final contradiction. □

Lemma 5.4.7. *Let S be non-degenerate and of finite rank, and R a linear* *subspace of S. Then every proper maximal linear subspace Q of R is a projective hyperplane of R.*

Proof. By lemma 5.4.6 there is a maximal linear subspace M of S such that $M \cap R = Q$. Let ℓ be a line of R such that $\ell \cap Q = \varnothing$. Let $p \in M \backslash Q$. As S is a polar space, p is collinear with some point q of ℓ. Thus $Q \cup \{p, q\}$ is a set of pairwise collinear points, and as $R = L(Q \cup \{q\})$ we see that $R \cup \{p\}$ is a set of pairwise collinear points for *every* $p \in M$. So $R \cup M$ is a set of pairwise collinear points and M maximal then implies $R \subseteq M$, a contradiction. Thus ℓ meets Q. □

Later we shall need the results of lemmas 5.4.8 and 5.4.9 below concerning linear subspaces and direct sums.

Lemma 5.4.8. *Let $L_{-1} \prec L_0 \prec L_1 \prec \ldots \prec L_{n-1}$ be a maximal chain of linear subspaces in a polar space S of finite rank n. If $S = \bigoplus S_i$, then the distinct elements of the chain $L_{-1} \cap S_i \subseteq L_0 \cap S_i \subseteq L_1 \cap S_i \subseteq \cdots \subseteq L_{n-1} \cap S_i$, for each i, form a maximal chain of linear subspaces of S_i.*

Proof. Suppose $L_j \cap S_i \subsetneq L_{j+1} \cap S_i$ for some j. If $L_j \cap S_i \subsetneq R \subsetneq L_{j+1} \cap S_i$ for some linear subspace R of S_i, then $L_j \subsetneq L_j \cup R \subsetneq L_{j+1}$ where $L_j \cup R$ is a linear subspace of S, contradicting $L_{-1} \prec L_0 \prec L_1 \cdots \prec L_{n-1}$ maximal. □

Lemma 5.4.9. *Let $S = \bigoplus S_i$, S a polar space of finite rank. If for each i, all maximal chains of S_i have the same number of elements, then all maximal chains of S have the same number of elements.*

Proof. Let $L_{-1} \prec L_0 \prec L_1 \ldots \prec L_{n-1}$ be a maximal chain of S. We count the number of elements of this chain in two ways. First, there are $n+1$ of them.

By lemma 5.4.8, for each i the chain $L_{-1} \cap S_i \subseteq L_0 \cap S_i \subseteq L_1 \cap S_i \subseteq \cdots \subseteq L_{n-1} \cap S_i$ gives rise to a maximal chain of linear subspaces of S_i with $n_i + 1$ elements, say, where n_i is independent of the maximal chain by assumption. But, by lemma 5.1.2, $L_j = \bigoplus_i (L_j \cap S_i)$. We count the number of *distinct* elements of $\bigoplus_i (L_j \cap S_i)$ for each j; there are $\Sigma(n_i + 1)$ of them. Hence $n + 1 = \Sigma(n_i + 1)$ is independent of the maximal flag chosen. □

5.5 Irreducibility

A polar space is called *reducible* if it is a direct sum of at least two non-empty polar spaces. Otherwise it is called *irreducible*.

The polar space of figure 5.1.2 (*c*) is clearly reducible. So is that of figure 5.1.2 (*b*) (check!) while that of figure 5.1.2 (*a*) is not.

Let us say that the points p and q of S are *adjacent*, written $p \sim q$, if they are collinear.

If $S = \bigoplus_{i \in I} S_i$ is a direct sum of polar spaces S_i, then each point in S_i is adjacent to all points of S_j for any $j \neq i$. Define a graph G as follows. The points of G are the points of S. The edges of G are the pairs (p, q) where p and q are distinct *non-adjacent* points. So, in G, a point $p \in S_i$ is never on an edge with a point $q \in S_j$ for any $j \neq i$.

A *component* of a graph G is a subgraph G' (i.e., a *subspace*, if we consider G as a near-linear space) of G such that no point of G' is adjacent to any point of $G \backslash G'$. G' is called a *connected component* (or a graph is called *connected*) if we can get from any point to another via a series of adjacent points.

It follows that, with the graph G of S using *non-adjacency* to define edges, each S_i is a component. Moreover, if S_i is irreducible, then S_i is connected. Lemma 5.5.1 tells us that every non-degenerate polar space can be written as a direct sum of irreducible polar spaces.

Lemma 5.5.1. *Every non-degenerate polar space is a direct sum of the connected components of the relation* $\not\sim$, *and these components are irreducible.*

Proof. Let S_i, i in some index set I, be the connected components of the relation $\not\sim$. Clearly then, the components S_i are irreducible under $\not\sim$. It suffices to show that each is a polar space and that any line joining a point of S_i to a point of S_j, $i \neq j$, consists of just two points.

To prove the latter, let $x \in S_i$, $y \in S_j$, $i \neq j$. Then $x \sim y$ implies the line xy exists. If xy contains a point $z \neq y$ not in S_i, then we may choose p to be a point not adjacent to x, since S is non-degenerate. Then p must be in S_i and $p \sim y$ and $p \sim z$ while this implies $p \sim x$ since S is a polar space. Contradiction!

Similarly, xy contains no point different from x in S_i, because we can use the above argument for y in S_j.

To see that S_i is a polar space we need only show that it is a subspace of S. Let x and y be points of S_i, $x \sim y$. Suppose there is a point z of xy not in S_i. Since S is non-degenerate, there is a point p not adjacent to z. Clearly, $p \notin S_i$. Thus $p \sim x$ and $p \sim y$ forcing $p \sim z$ and a final contradiction. □

Corollary. *A non-degenerate polar space is irreducible if and only if the graph of the relation* $\not\sim$ *is connected.*

Proof. If S is not connected then, as in the proof of lemma 5.5.1, S is reducible. Conversely, if S is a direct sum of polar spaces S_i, it is easily seen that points in distinct sets S_i cannot be in the same connected component of $\not\sim$. □

Lemma 5.5.2. *Any polar space has at most one representation as a direct sum of irreducible polar spaces.*

Proof. Let $S = \bigoplus S_i = \bigoplus T_i$ for irreducible polar spaces S_i and T_i. For a fixed

S_i we may suppose that $S_i \cap T_j \neq \emptyset$ for some T_j. Then $S_i = (S_i \cap T_j) \oplus S_i \setminus T_j$ contradicting S_i irreducible unless $S_i \setminus T_j = \emptyset$, in which case $S_i \subseteq T_j$. But if $S_i \subseteq T_j$, $T_j = S_i \oplus (T_j \setminus S_i)$, contradicting T_j irreducible. So $S_i = T_j$.

Thus each factor S_i is a T_j and conversely, so that the representations of S are the same except perhaps for the order of the factors. □

Lemmas 5.5.1 and 5.5.2 tell us that any non-degenerate polar space has a unique representation as a direct sum of irreducible polar spaces.

Lemma 5.5.3. *In any polar space S with non-adjacent points p and q, $p^\perp \cap q^\perp$ is a subspace of S (and so a polar space).*

Proof. We leave this to the reader. □

Lemma 5.5.4. *Let S be an irreducible polar space of finite rank. Let $p \not\sim q$ and $\bar{p} \not\sim \bar{q}$ be points of S. Then the polar spaces $p^\perp \cap q^\perp$ and $\bar{p}^\perp \cap \bar{q}^\perp$ are isomorphic. (We write $p^\perp \cap q^\perp \approx \bar{p}^\perp \cap \bar{q}^\perp$.)*

Proof. We must exhibit a 1–1 onto linear function from $p^\perp \cap q^\perp$ to $\bar{p}^\perp \cap \bar{q}^\perp$. Suppose first of all that $p = \bar{p}$. For any point x of $p^\perp \cap q^\perp$, the line px contains a unique point x' say, adjacent to $\bar{q}$. We map x to x'. It is easy to see that this is 1–1 and onto. It also preserves adjacency and non-adjacency because of the following argument. Suppose x and y are adjacent points of $p^\perp \cap q^\perp$. Then $x \sim y$ and $x \sim p$ implies $x \sim y'$, $y' \in py$. Thus, $y' \sim x$ and $y' \sim p$ implies $y' \sim x' \in px$. Similarly, $x' \sim y'$ implies $x \sim y$. So maximal linear subspaces of $p^\perp \cap q^\perp$ are mapped to maximal linear subspaces of $\bar{p}^\perp \cap \bar{q}^\perp$.

By lemma 5.4.6 each line is an intersection of maximal linear subspaces and so lines are mapped to lines.

To complete the proof it remains only to note that since S is irreducible, and therefore connected in the graph-theoretical sense, we can always find a chain of points $p = p_1, p_2, \ldots, p_m = \bar{p}$ such that $p_i \not\sim p_{i+1}$ (see the corollary to lemma 5.5.1). Then, by the first part of the proof, $p^\perp \cap q^\perp = p_1^\perp \cap q^\perp \approx p_1^\perp \cap p_2^\perp \approx p_2^\perp \cap p_3^\perp \approx \cdots \approx p_{m-1}^\perp \cap p_m^\perp \approx p_m^\perp \cap \bar{q}^\perp = \bar{p}^\perp \cap \bar{q}^\perp$. □

5.6 Projective spaces inside polar spaces

A *generalized projective plane* is a linear space which satisfies PP1 but not necessarily PP2. A *generalized projective space* is a linear space in which every plane is a generalized projective plane. (See exercise 39 of section 5.8.)

In this section we present the important result that each linear subspace of a non-degenerate polar space of finite rank is a generalized projective space. (The reader should examine the linear subspaces of the polar spaces of figures 5.1.1 and 5.1.2.)

Lemma 5.6.1. *Let S be a non-degenerate polar space of finite rank. Let M be* *a maximal linear subspace of S and R a proper maximal linear subspace of M. Then there are non-adjacent points p and q such that* $R \subseteq p^\perp \cap q^\perp$ *and R is a maximal linear subspace of* $p^\perp \cap q^\perp$.

Proof. By lemma 5.4.6 there is a maximal linear subspace M' such that $M \cap M' = R$. Also, by lemma 5.4.7, R is a projective hyperplane of M. Let $p \in M \setminus R$ and $q \in M' \setminus R$. Since R is a hyperplane (lemma 2.5.6) of M, $M = L(R \cup \{p\})$. If $p \sim q$, then $q^\perp \supseteq R \cup \{p\}$. If $q \in M$, then $M = L(R \cup \{q\}) \subseteq M \cap M' = R$, a contradiction, so we may now apply lemma 5.4.2 to get $q^\perp \supseteq L(R \cup \{p\}) = M$ and so $M \cup \{q\}$ is a clique, contradicting the maximality of M. Therefore, $p \not\sim q$. Clearly $R \subseteq p^\perp \cap q^\perp$. Suppose that $R \subsetneq R' \subseteq p^\perp \cap q^\perp$, R' a maximal linear subspace of $p^\perp \cap q^\perp$. Clearly $p \in R'$. Then $L(R' \cup \{p\})$ is a linear space containing M, so that the maximality of M implies $R' = M$, while $M \not\subseteq q^\perp$. Hence R is maximal in $p^\perp \cap q^\perp$. □

Lemma 5.6.2. *Let S be a non-degenerate polar space of finite rank n. Then all maximal chains of linear subspaces have* $n+1$ *elements.*

Proof. We proceed by induction on n. $\emptyset$ and any singleton are degenerate, so the first case to examine is $n = 2$. In this case, each point is on a line so that the lemma holds true.

Suppose then that $n > 2$ and that there are distinct maximal chains in S, $L_{-1} < L_0 < L_1 < \cdots < L_{n-1}$ and $\bar{L}_{-1} < \bar{L}_0 < \bar{L}_1 < \cdots < \bar{L}_{m-1}$ where $n \neq m$. By lemma 5.6.1 there are points p, q, r and s, $p \not\sim q$, $r \not\sim s$ such that $L_{n-2} \subseteq p^\perp \cap q^\perp$ and $\bar{L}_{m-2} \subseteq r^\perp \cap s^\perp$, and such that L_{n-2} and $\bar{L}_{m-2}$ are maximal linear subspaces in the polar spaces $p^\perp \cap q^\perp$ and $r^\perp \cap s^\perp$ respectively. By induction, all maximal chains of linear subspaces of these latter polar spaces have length n and m respectively. It follows that $p^\perp \cap q^\perp$ and $r^\perp \cap s^\perp$ are not isomorphic. Because of lemma 5.5.4, S must be reducible. Let $S = S_1 \oplus S_2$, S_1 and S_2 non-empty polar spaces. As S is non-degenerate, so are S_1 and S_2 (see exercise 3(b) of section 5.8). Clearly the rank of each S_i is less than n and so by induction all maximal chains of linear spaces of S_i have the same length, $i = 1, 2$.

We obtain a contradiction now from lemma 5.4.9. □

Corollary. *Let S be a non-degenerate polar space of finite rank n. Write* $S = \bigoplus S_i$. *Then* $n = \sum \operatorname{rank} S_i$.

Proof. This is immediate from lemma 5.6.2 and the proof of lemma 5.4.9.

Lemma 5.6.3. *Let S be a non-degenerate polar space of finite rank. Then every linear subspace of S is a generalized projective space.*

Proof. 'Planes' are linear subspaces of rank 3. We must prove that in each plane any two lines meet. By lemma 5.6.2, any line is a proper maximal linear subspace of a plane. By lemma 5.4.7, such a proper maximal linear subspace is a projective hyperplane and hence meets any other line. □

5.7 A history of polar spaces

Because the theory of polar spaces is relatively new, it would be a good idea before leaving this chapter to consider the motivation for their study and the resulting development of the theory.

It was recognized that the set of absolute points of a (non-degenerate) polarity in a Desarguesian projective space had many properties in common with those of non-degenerate quadrics in projective space over a field. The question raised was, is it possible to find a structure that includes both of these systems, but not too much more? Veldkamp (1959) was the first to come up with such a system. There were problems with his theory in the characteristic 2 case, however. Tits (1974) improved on Veldkamp's system by coming up with the following definition.

A *polar space* S is a set of points together with distinguished subsets called subspaces such that

(i) a subspace together with the subspaces it contains is an i-dimensional projective space with $-1 < i \leq n-1$ for some integer n (the smallest n for which this occurs is called the *rank* of S);
(ii) the intersection of any two subspaces is a subspace;
(iii) given a subspace R of dimension $n-1$ and a point $p \in S \setminus R$ there exists a unique subspace M containing p such that $\dim(M \cap R) = n-2$. M contains all points of R which are joined to p by some subspace of dimension 1 (a line);
(iv) there exist disjoint subspaces of dimension $n-1$.

(Some of these properties should sound familiar. The reader is referred to lemmas 5.4.4 and 5.4.5.)

Tits (1974) was also able to give a complete classification of polar spaces as long as the rank was at least 3:

Theorem 5.7.1. *Let S be a polar space of rank greater than or equal to 3. Then one and only one of the following situations is realized.*

(1) *S is a polar space arising from a polarity in a projective space determined by a trace-valued σ-Hermitian form.*[†]
(2) *S is a polar space arising from a pseudo-quadratic form on a division*

[†] See Tits (1974) or Hughes and Piper (1973).

ring F[†] *in a projective space, where F has characteristic 2 and the antiautomorphism σ involved satisfies $\sigma^2 = 1$ and $\{t \in F \mid t^n = t\} \neq \{u + \sigma(u) \mid u \in F\}$.*

(3) *S is a polar space arising from a symplectic polarity*[‡] *in a projective space over a field of characteristic $\neq 2$.*

(4) *S is a polar space of rank 3 whose maximal subspaces are Moufang planes.*[§]

(5) *S is a polar space of rank 3 corresponding to a three-dimensional projective space on a non-commutative division ring.*

The polar spaces of rank 2 are generalized quadrangles. We shall consider them in more detail in chapter 6.

It is clear from this classification theorem that the structures studied in sections 5.2 and 5.3 by no means cover the family of all polar spaces.

Ernest Shult independently studied the characterization problem over the field GF(2) (Shult, 1972). His results were framed in graph-theoretic language and the main theorem is as follows

Theorem 5.7.2. *Let G be a regular finite graph. Assume that for every edge (a, b) of G there exists a point c such that* (i) *(a, c) and (b, c) are edges,* (ii) *each point of $G \backslash \{a, b, c\}$ is adjacent to either one or all points of $\{a, b, c\}$. Then unless G is complete or has no edges it is isomorphic to one of the systems* (1) *or* (2) *of theorem 5.7.1.*

It should be clear to the reader that the properties of G in theorem 5.7.2. are very close to those properties we use to define a polar space in section 5.1.

The graph-theoretic approach became very useful as soon as it was suggested that maximal sets of pairwise adjacent points in the graph satisfying the 'one-all' property described in (ii) of theorem 5.7.2 be considered as lines. The concept of a *Shult space* was introduced by Buekenhout and Shult (1974).

A *Shult space* S is a set of points together with distinguished subsets of cardinality greater than or equal to 2 called *lines* such that for each line ℓ of S and each point $p \in S \backslash \ell$, p is adjacent to either one or all points of ℓ. (Two points are *adjacent* if there is *at least one* line of S containing them.)

A Shult space is *non-degenerate* if no point is adjacent to all other points.

A *subspace* R of a Shult space S is a non-empty set of pairwise adjacent points such that any line meeting R in more than one point is contained

[†] See Tits (1974).
[‡] See Dembowski (1968).
[§] See Kallaher (1982).

in R. S is of *finite rank* if there is an integer n such that each chain of distinct subspaces $R_1 \subsetneq R_2 \subsetneq \cdots \subsetneq R_i$ of S has at most n elements.

The main result of Buekenhout and Shult (1974) is then the following.

Theorem 5.7.3. *Let S be a non-degenerate Shult space of finite rank all of whose lines have cardinality greater than or equal to 3. Then S is a polar space.*

The definition of polar space used in this chapter is due to the next result found in Buekenhout and Shult (1974).

Theorem 5.7.4. *Every polar space is a Shult space.*

5.8 Exercises

1. Find new examples of non-degenerate polar spaces.
2. Classify those polar spaces with not more than six points.
3. (*a*) If S is a direct sum of polar spaces, none of which is degenerate, show that S itself is not degenerate.
   (*b*) If S is a direct sum of polar spaces at least one of which is degenerate, show that S itself is degenerate.
4. Show that any subspace of a polar space is a polar space.
5. Let $\{S_i\}$, $i \in I$, be a family of polar spaces and let $\{E_i\}$, $i \in I$, be a family of linear subspaces, $E_i \subseteq S_i$. Show that the direct sum E of the E_i is a linear subspace of the direct sum S of the S_i.
6. Find a polarity of the projective plane of order 3 and compute the absolute points.
7. Show that if Π is a projective plane of order 7 then Π cannot have precisely 29 absolute points.
8. Show that if the projective plane Π has order 5, then the number of absolute points of any given polarity on it is a multiple of 3. (Use theorem 5.2.3.)
9. If a projective plane of order 10 did exist and had a polarity Π, show that the number of absolute points of Π would be congruent to 11 modulo 49.
10. Show that every polarity of a plane of finite order has an absolute point. (Use theorem 5.2.3.)
11. Show that if R is a totally isotropic subspace of the projective space S of dimension d, then $\dim R \leq (d-1)/2$.
12. In $P(2,7)$ define $\sigma: [p_0, p_1, p_2] \leftrightarrow \Sigma_{i=0}^{2} p_i x_i = 0$. Find the absolute points of σ.
13. In $P(3,5)$ define $\sigma: [p_0, p_1, p_2, p_3] \leftrightarrow \Sigma_{i=0}^{3} p_i x_i = 0$. Find the absolute points and non-isotropic lines.

14. Show that an isotropic line of a polarity in a projective space of finite dimension greater than 2 is not an absolute line.

15. Define an *ovoid* O in a projective space S to be a non-empty set of points satisfying $|O \cap \ell| \le 2$ for all lines ℓ, and such that for each point p of O, the union of all points on all lines tangent to O at p is the set of points of a hyperplane. Construct an example. (Hint: see questions 16, 17 and 18. Try dimension 2 first.)

16. Show that, if O is an ovoid of a projective space $P(d, k)$, then $|O| = k^{d-1}+1$.

17. Show that if H is a hyperplane of any projective space S, and O is an ovoid of S then, if $O \cap H \ne \varnothing$, $O \cap H$ is an ovoid of H.

18. Prove that if the projective space $P(d, k)$ contains an ovoid then $d \le 3$. (Hint: use exercises 16 and 17 and prove the result for $P(4, k)$.)

***19.** If k is odd show that every ovoid in $P(d, k)$ where $d = 2$ or 3 is a non-degenerate quadric. What is its rank?

20. Determine the intersection of the quadric Q and the line pq in projective 3-space for the following:
(*a*) $F =$ reals, $Q: x_0^2 + x_2^2 + x_2x_3 = 0$, $p = [1, 0, 1, 1]$, $q = [0, 1, 1, 1]$;
(*b*) $F =$ the complex numbers, $Q: x_0^2 + x_1^2 - x_2x_3 = 0$, $p = [1, 1, 1, 1]$, $q = [1, 2, 3, 4]$;
(*c*) $F = \mathrm{GF}(2)$, $Q: x_0x_1 + x_1x_2 + x_3^2 = 0$, $p = [1, 0, 1, 0]$, $q = [0, 1, 1, 0]$.

21. Compute the matrix of the quadric $x_1^2 + x_0x_1 + x_1x_3 + x_3^2 = 0$ in $P(3, 3)$. Find some pairs of distinct conjugate points.

22. Determine all points on the quadric of example 5.3.1.

23. Determine all points on the quadric $x_0^2 + x_1^2 + x_2^2 = 0$ in two-dimensional projective space (*a*) over the field of real numbers, (*b*) over the field of complex numbers.

24. A *singular point* of a quadric Q with matrix A is a point p satisfying $pA = 0$. Find all singular points of the quadric of example 5.3.3.

25. Give examples of ruled and unruled quadrics.

26. A *conic* is a quadric in projective 2-space. Show that if a conic is reducible (i.e., its equation is factorable), then it consists of either one or two lines. Show moreover that, if it consists of one line only, then every point is singular while, if it consists of two lines, only the point of intersection is singular.

27. Let C be a conic in the projective plane Π. (See exercise 26.) If C is not reducible and is not a projective subspace of Π, show that every line of Π meets it in 0, 1, or 2 points and each point of C is on a unique tangent to C. (C is called a *conic oval*.)

28. Let $xAx^t = 0$ be the equation of a quadric. For any point p with corresponding vector $p = [p_0, p_1, \ldots, p_d]$ define $\sigma(p) = [pAx^t = 0]$. That is,

σ takes p to the set of points satisfying $pAx^t = 0$. Show that σ is a polarity.

29. Let S be a projective space and H a hyperplane. Let $p \in S \setminus H$. Determine H_p. (See page 100.)

30. Repeat exercise 29 in case S is an affine space.

31. Classify all polar spaces of rank 1.

32. Find a polar space of infinite rank.

33. Let S be a polar space and a and b non-collinear points. Show that $\operatorname{rad}(a^\perp \cap b^\perp) \subseteq \operatorname{rad} a^\perp$.

34. If $S = \bigoplus S_i$ is a direct sum of polar spaces, show that $\operatorname{rad} S = \bigcup \operatorname{rad} S_i$.

35. Draw the adjacency and non-adjacency graphs of the polar spaces of figure 5.1.2. (So the points of G are the points of the polar space and (x, y) is an edge when $x \sim y$ or, respectively, when $x \nsim y$. See lemma 5.5.1.)

36. Show that the function L of section 5.4 makes S into a closure space.

37. Is lemma 5.5.1 true for degenerate polar spaces? (Explain!)

38. Show that lemma 5.5.4 can be false in the reducible case.

39. Show that the generalized projective planes are the projective planes and the near-pencils. (See the de Bruijn–Erdös theorem, 2.2.2.)

40. In any polar space with non-adjacent points p and q show that $p^\perp \cap q^\perp$ is a subspace.

41. Find the rank, as a polar space, of the affine space $A(d, k)$.

42. When is the R_p of lemma 5.4.3 a blocking set in the sense of chapter 8?

6

Generalized quadrangles

For everything there is a season, and a time for every matter under heaven...
a time to break down and a time to build up;...
Ecclesiastes 3, verse 3

6.1 Definition and some basic results

We continue to use the words 'collinear' and 'adjacent' interchangeably. A *generalized quadrangle* is a near-linear space $S = (P, L)$ satisfying

GQ1 for each point p and line ℓ, $p \notin \ell$ implies $c(p, \ell) = 1$;

GQ2 there exist non-collinear points and non-concurrent lines;

GQ3 P is a finite set.

Notice that a generalized quadrangle is a polar space. Notice also that, like the projective plane axioms, the generalized quadrangle axioms have 'duality'. It follows that the dual space of a generalized quadrangle is again a generalized quadrangle.

Example 6.1.1. A generalized quadrangle in which every point is on two lines is called a *grid*. (See figure 6.1.1.)

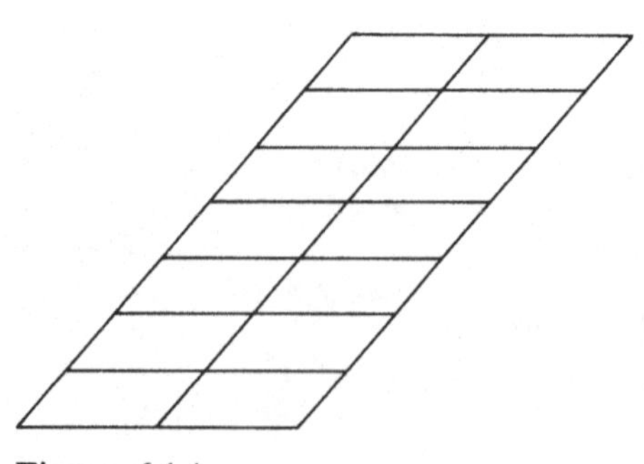

Figure 6.1.1.

It is possible to show that, if each point is on two lines, then the generalized quadrangle is like that of figure 6.1.1 in the sense that the lines are partitioned into two sets L_1 and L_2 and each line of L_1 has a fixed number of points s_1, say, while each line of L_2 has a fixed number of points s_2.

Example 6.1.2. The dual of figure 6.1.1 has eleven points partitioned into two sets of three and eight elements. Any generalization of this given by a partition of a point set P into sets P_1 and P_2 such that each element of P_1 is joined to each element of P_2, while no element of P_i is joined to an element of P_i, $i = 1, 2$, is clearly also a generalized quadrangle. It is called a *dual grid.*

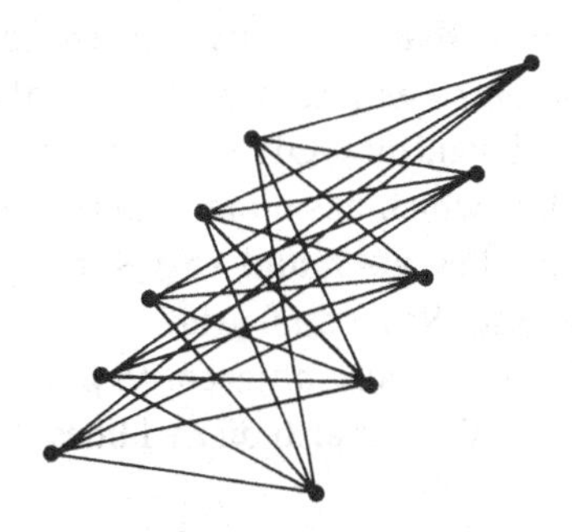

Figure 6.1.2.

Example 6.1.3. Let $A = \{1,2,3,4,5,6\}$. A *duad* will be a subset of A containing two elements. A *syntheme* will be a triple of disjoint duads. The reader should check that there are fifteen duads and fifteen synthemes. Now let $S = (P, L)$ where P is the set of duads and L the set of synthemes. It is easily checked that GQ1 and GQ2 hold. For example, if $p = \{1,2\}$ and $\ell = \{\{1,3\},\{2,4\},\{5,6\}\}$, then the unique syntheme on p meeting ℓ is $\{\{1,2\}\{3,4\},\{5,6\}\}$.

The words 'duad' and 'syntheme' were used by Sylvester (1861, 1884) who first introduced this particular geometry.

We shall show that the above is the only example of a generalized quadrangle on fifteen points and fifteen lines. To see this, the next theorem and lemmas are very useful.

Theorem 6.1.1. *Let $S = (P, L)$ be a generalized quadrangle. Then one of the following is true:*

(i) *S is a grid ($P \neq \varnothing$) with each point on two lines and the set of lines partitioned into two sets, L_1 and L_2, each line meeting every line of the other set and no line of the set containing it. Moreover, lines of L_i have the same constant number of points s_i, $i = 1, 2$.*

(ii) *S is a dual grid.*

(iii) *There are integers $s \geq 1$ and $t \geq 1$ such that each line is on $s+1$ points and each point is on $t+1$ lines.*

Proof. The fact that any point is on at least two lines follows easily from GQ2 and GQ1. We prove then that any two non-adjacent points are on the same number of lines. Let p be on $t+1$ lines and $q \nsim p$. By GQ1, q is on at least $t+1$ lines. If q were on more than $t+1$ lines then, by GQ1, p would be on more than $t+1$ lines.

Now let $p \sim q$. Not all points are adjacent to p because of GQ2. Let $r \nsim p$. Then r is on $t+1 \geq 2$ lines. If $r \nsim q$, then q is also on $t+1$ lines and we are done. So suppose $r \sim q$. Suppose that all points adjacent to r are also adjacent to p. Hence, each line on r has just two points. Label the points adjacent to r as $x_1, x_2, \ldots, x_n$. Then, by GQ1, $x \nsim r$ implies $x \sim x_i$ for all i. If any such point x is on a line which has more than two points, letting $y \in xx_j \setminus \{x, x_j\}$ for some x_j implies $y \nsim r$ and so $y \sim x_i$, $i = 1, 2, \ldots, n$, contradicting GQ1. So in this case, S is a dual grid.

If there is a point x adjacent to r and not also adjacent to p, then x is on $t+1$ lines. But $x \nsim q$ (since $r \sim q$) implies q is also on $t+1$ lines. Hence, all points are on $t+1 \geq 2$ lines.

The dual of the above argument implies that either S is a grid or all lines have the same number $s+1 \geq 2$ of points. □

Note that, if S is a grid with $s_1 = s_2$, then S falls into case (iii) also. If S is a dual grid with points partitioned into the sets P_1 and P_2, and if $|P_1| = |P_2|$, then S falls into case (iii) as well.

If S is of type (iii) we shall say it has *parameters s and t*. We shall still use the letter t ($=1$) for a grid, and s ($=1$) for a dual grid, however.

Generalized quadrangles have some very nice combinatorial properties and we shall give some of the more elementary ones in the remainder of this section.

Lemma 6.1.2. *Let p and q be non-collinear points. Then there are precisely $t+1$ points collinear with both p and q.*

Proof. By GQ1, q is collinear with a unique point on each of the $t+1$ lines on p. Clearly this gives us precisely $t+1$ points collinear with both p and q. □

Lemma 6.1.3. *In any generalized quadrangle with parameters s and t, $v = (s+1)(st+1)$ and $b = (t+1)(st+1)$.*

Proof. Fix a line ℓ. There are $v-(s+1)$ points not on ℓ. Each point not on ℓ is on a unique line meeting ℓ. Hence, an alternative way to count all the points not on ℓ is to count all the points on lines meeting ℓ. There are $(s+1)t$ such lines (using theorem 6.1.1) and each has s points not on ℓ.

Hence, $v-(s+1)=st(s+1)$ or $v=st(s+1)+(s+1)=(s+1)(st+1)$. Dually, $b=(t+1)(st+1)$. □

With the information from the above results, we can now show fairly easily that the generalized quadrangle of example 6.1.3 is unique. Let $P=\{1,2,3,\ldots,15\}$, $v=b=15$. Because of lemma 6.1.3, $s>1$ and $t>1$. Hence S falls into category (iii) of theorem 6.1.1. In fact, the result of lemma 6.1.3 implies $s=t=2$, so that each line has three points and each point is on three lines. We may, without loss of generality, let $\{1,2,3\}$ be a line. Each point 1 and 3 must be on other lines which cannot meet by GQ1. Let $\{3,4,5\}$ and $\{1,9,10\}$ be such lines. By GQ1, the point 9 is on a line meeting $\{3,4,5\}$, but there is a third line on 9 which does not meet any of the given lines so far. Let $\{7,8,9\}$ be this line. Now the point 5 must meet $\{7,8,9\}$ in a line. Again, without loss of generality, we may suppose that this line is $\{5,6,7\}$. (See figure 6.1.3.)

Because of GQ1 there must be lines on $\{1,6\}$, on $\{3,8\}$, on $\{5,10\}$, on $\{7,2\}$ and on $\{9,4\}$. Also the point 2, which is so far only on two lines, must be on lines meeting the lines on $\{4,9\}$ and on $\{5,10\}$, by GQ1. Since 2 is on precisely three lines, we must have the same line on 2 meeting both lines $\{4,9\}$ and $\{5,10\}$. This also means, by GQ1, that we have found the third points on the lines on $\{4,9\}$ and $\{5,10\}$. Let them be 11 and 12 respectively. So $\{2,11,12\}$, $\{4,9,11\}$ and $\{5,10,12\}$ are lines. A

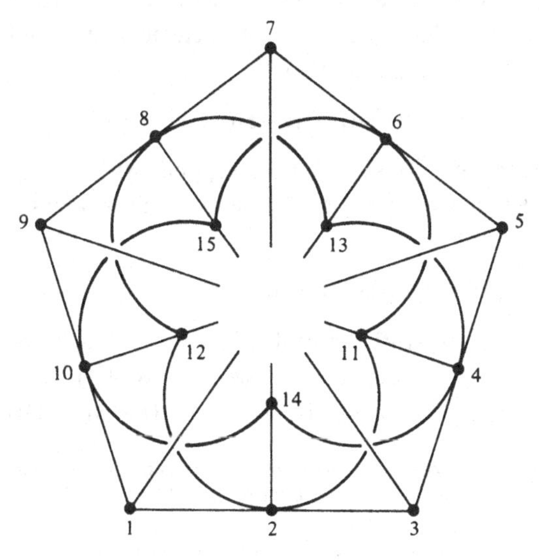

Figure 6.1.3.

similar argument for the point 4 with respect to the lines on $\{1,6\}$ and $\{2,7\}$ leads to lines $\{4,13,14\}$, $\{1,6,13\}$ and $\{2,7,14\}$, say.

Some line on 6 must meet $\{4,9,11\}$. Since $\{4,9,11\}$ already has three points, GQ1 implies that 6 and 11 are on a common line. Also, using GQ1 for 6 and the line on $\{3,8\}$, we see that we have a point 15, say, such that $\{3,8,15\}$ and $\{6,11,15\}$ are lines.

Final arguments with the points 8 and 15 show that $\{8,12,13\}$ and $\{10,14,15\}$ are lines. We now have fifteen lines, and GQ1, GQ2 and GQ3 are satisfied.

Our construction was done completely without loss of generality, so this is the unique generalized quadrangle with $v = b = 15$. We leave it to the reader (exercise 2 of section 6.8) to find the isomorphism between figure 6.1.3 and the system of duads and synthemes presented earlier.

Lemma 6.1.4. *In any generalized quadrangle S with parameters s and t,* $(s+t)\,|\,st(s+1)(t+1)$.[†]

Proof. List the points of S as $\{p_1, p_2, \ldots, p_v\}$. Let A be the matrix (a_{ij}) where $a_{ij} = 1$ if $i \neq j$ and $p_i \sim p_j$, and $a_{ij} = 0$ otherwise, $1 \le i, j \le v$. Consider the matrix A^2 with entries c_{ij} say, $1 \le i, j \le v$. Then $c_{ii} = (t+1)s$, $1 \le i \le v$. If $i \neq j$ and $p_i \nsim p_j$, then $c_{ij} = (t+1)$, $1 \le i, j \le v$. If $i \neq j$ and $p_i \sim p_j$, then $c_{ij} = s-1$, $1 \le i, j \le v$. It follows that

$$A^2 - (s-t-2)A - (t+1)(s-1)I = (t+1)J$$

where I is the $v \times v$ identity matrix with 1s on the diagonal and 0 everywhere else, and J is the $v \times v$ matrix with all entries equal to 1.

So $(t+1)s$ is an eigenvalue[‡] of A. But J has eigenvalues 0 and v with respective multiplicities $v-1$ and 1. Also

$$((t+1)s)^2 - (s-t-2)(t+1)s - (t+1)(s-1) = (t+1)v$$

implies that the eigenvalue $s(t+1)$ of A corresponds with the eigenvalue v of J. Hence $s(t+1)$ has multiplicity 1. The other eigenvalues of A are roots of

$$x^2 - (s-t-2)x - (t+1)(s-1) = 0.$$

Let the roots be r_1 and r_2 with multiplicities m_1 and m_2 respectively. Then $r_1 = -t-1$, $r_2 = s-1$ and $1 + m_1 + m_2 = v$, $s(t+1) - m_1(t+1) + m_2(s-1) = \text{trace of } A = 0$. Therefore, $s(t+1) - (v - m_2 - 1)(t+1) +$

† Recall that $a\,|\,b$ for integers a and b means that b is a multiple of a. The notation is usually read as 'a divides b'. The result presented here was first due to Bose (1972).

‡ We need some results from linear algebra in the remainder of the proof. The reader is referred to Curtis (1967).

$m_2(s-1)=0$ implies

$$m_2 = \frac{(v-1)(t+1)-s(t+1)}{s+t}$$

$$= \frac{st(s+1)(t+1)}{s+t} \text{ which must be an integer. } \square$$

Before proving the next theorem, we need a result about inequalities. Let $t_1, t_2, \ldots, t_d$ be real numbers and d a positive integer. Then

Lemma 6.1.5. $d\sum_{i=1}^{d} t_i^2 \geq (\sum_{i=1}^{d} t_i)^2$.

Proof. Set $\bar{t} = (\sum_{i=1}^{d} t_i)/d$. Then

$$0 \leq \sum_{i=1}^{d} (\bar{t}-t_i)^2$$

$$= \sum_{i=1}^{d} (\bar{t}^2 - 2\bar{t}t_i + t_i^2) = d\bar{t}^2 - 2\bar{t}\sum_{i=1}^{d} t_i + \sum_{i=1}^{d} t_i^2$$

$$= \left(\sum_{i=1}^{d} t_i\right)^2 \Big/ d - 2\left(\sum_{i=1}^{d} t_i\right)^2 \Big/ d + \sum_{i=1}^{d} t_i^2.$$

So $$d\sum_{i=1}^{d} t_i^2 \geq \left(\sum_{i=1}^{d} t_i\right)^2. \quad \square$$

Theorem 6.1.6. *If $s > 1$ then $t \leq s^2$. Dually, if $t > 1$ then $s \leq t^2$.*[†]

Proof. Let p and q be non-collinear points. Define $X = \{x | p \not\sim x \text{ and } q \not\sim x\}$. So $|X| = v - 2 - 2(t+1)s + (t+1)$ (by lemma 6.1.2) $= (s+1)(st+1) - 2 - (2s-1)(t+1)$ by lemma 6.1.3. For each $x_i \in X$, let $t_i = |\{y | y \sim p, y \sim q \text{ and } y \sim x_i\}|$, $1 \leq i \leq |X| = d$.

Now count in two different ways the number of ordered pairs (x_i, y), $x_i \in X$, $y \sim p, q, x_i$. We obtain $\sum_i t_i = (t+1)(t-1)s$ where the right hand side is the number of points not on lines on p or q but on lines containing points joined to p and q. (We use lemma 6.1.2 here once again.) Next, count in two different ways the number of ordered triples (x_i, y, y') where y and y' both satisfy the conditions placed on y above, and $y \neq y'$. One way of counting is given by $\sum_i t_i(t_i - 1)$ since $y \neq y'$. On the other hand, the number of ways of choosing y and y' is $(t+1)t$ by lemma 6.1.2. Since there are $t+1$ lines on y, but two of these are yp and yq, it follows that y' meets only $t-1$ lines on y in points not collinear with p or q. So $\sum_i t_i(t_i-1) = (t+1)t(t-1)$.

Combining both equations yields $\sum_i t_i^2 = (t+1)(t-1)(s+t)$. Since $d\sum_i t_i^2 - (\sum_i t_i)^2 \geq 0$ by lemma 6.1.5, we obtain $d(t+1)(t-1)(s+t) \geq$

[†] This result was first due to Higman (1971) while the proof given is from P. Cameron (1974).

$(t+1)^2(t-1)^2s^2$ or $d(s+t) \geq (t^2-1)s^2$. Consequently, $t(s-1)(s^2-t) \geq 0$ or $s^2 \geq t$. □

If S is a generalized quadrangle with parameters $s=2$ and $t>1$, then, by theorem 6.1.6, $t \leq s^2 = 4$. Also, lemma 6.1.4 implies $t=2$ or 4. In either case, there is a unique quadrangle. In fact the quadrangle with $t=2$ is exhibited in figure 6.1.3. We shall be considering the case $s=2$, $t=4$ in a more general context in the next section, and shall take a special look at the case $s=t=3$ in section 6.4.

6.2 All known examples

In this section we examine the known classes of generalized quadrangles. The first three introduced below are known as the 'classical' examples and are all due to J. Tits.

Recall that theorem 5.3.3 says that quadrics over fields of characteristic not 2 are polar spaces. If the rank of the polar space is 2, it is easy to see that we have generalized quadrangles in the non-degenerate case. Consider, then, a non-singular quadric of rank 2 in the projective space $P(d,k)$. Hence $d=3$ and the quadric is hyperbolic, $d=5$ and the quadric is elliptic, or $d=4$.

Using the formula given in section 5.3 for counting m-dimensional subspaces, we find that

in $P(3,k)$: $s=k$, $v=(s+1)^2$, $b=2(s+1)$ and so $t=1$. (S is a grid.)
in $P(4,k)$: $s=k$, $v=(s+1)(s^2+1)=b$ and so $t=s$.
in $P(5,k)$: $s=k$, $v=(s+1)(s^3+1)$, $b=(s^2+1)(s^3+1)$ and so $t=s^2$.

We use the notation $Q(d,k)$ for the above examples.

As we saw in exercise 28 of section 5.8, a quadric can define a polarity. On the other hand, any polarity σ of a projective space over a field is connected with a non-singular matrix and a field automorphism α. If this matrix A is symmetric and $\alpha=1$, we have the quadric case. If A satisfies $\alpha(A^t)=A, \alpha^2=1$ and $\alpha \neq 1$, we call the polarity *Hermitian* or *unitary*. Such polarities exist in $P(d,k)$ if and only if k is a square q^2. By theorem 5.2.5, the set of absolute points of σ equipped with the totally isotropic lines of σ is a polar space. The number of m-dimensional subspaces of $P(d,q^2)$ in this polar space is given by

$$\prod_{i=d-2m}^{d+1} (q^i-(-1)^i) \Big/ \prod_{j=1}^{m+1} (q^{2j}-1).$$

(See exercise 46 of section 6.8.)

If S is to be a generalized quadrangle, the rank must be 2 which implies $d=3$ or 4. Computing the number of points from the above formula yields

in $P(3,q^2)$: $s=q^2$, $v=(q^2+1)(q^3+1)$, $b=(q+1)(q^3+1)$ and so $t=q$;

in $P(4,q^2)$: $s=q^2$, $v=(q^2+1)(q^5+1)$, $b=(q^3+1)(q^5+1)$ and so $t=q^3$.

We use $H(d,q^2)$ to denote these generalized quadrangles.

If the above matrix A is not symmetric, each diagonal entry is zero and $\alpha=1$, the polarity is called *symplectic* or *null*. $P(d,k)$ admits symplectic polarities if and only if d is odd. The number of m-dimensional totally isotropic subspaces is $\prod_{i=0}^{m}(k^{d+1-2i}-1)/(k^{i+1}-1)$.

If S is to have rank 2, then $d=3$. We denote this quadrangle by $W(k)$ with

$$s=k,\ v=(k+1)(k^2+1)=b \text{ and so } t=k.$$

Notice that the parameters of $W(k)$ are the same as those of $Q(4,k)$. In fact, it is known[†] that $W(k)$ is isomorphic to the dual of $Q(4,k)$. Also, $Q(5,k)$ is isomorphic to the dual of $H(3,k^2)$[‡] and $W(k)$ is self-dual if and only if k is even.[§]

Example 6.2.1. To illustrate the constructions mentioned above, consider the matrix

$$A=\begin{pmatrix} 0 & 1 & 0 & 0\\ -1 & 0 & 0 & 0\\ 0 & 0 & 0 & 1\\ 0 & 0 & -1 & 0 \end{pmatrix}$$

which induces a symplectic polarity in $P(3,k)$. If $k=2$, the equation satisfied by an absolute point $[x_0,x_1,x_2,x_3]$ is $-x_1x_0+x_0x_1-x_3x_2+x_2x_3=0$, which holds for any choice of x_0,x_1,x_2,x_3 not all zero. hence there are 15 points (all the points of $P(3,2)$) in the generalized quadrangle.

There are four 'non-classical' constructions of generalized quadrangles. The first is constructed as follows. Let O denote a set of $1+k^{d-1}$ points of $P(d,k)$, no three of which are collinear, $d=2,3$. O is an *oval* or *ovoid* according as $d=2$ or 3. (See exercises 15 to 19 of section 5.8.) Embed $P(d,k)$ as a hyperplane in $P(d+1,k)$. Define points of a new structure $T(O)$ to be of the following three types: (i) points of $P(d+1,k)$ not in $P(d,k)$, (ii) hyperplanes of $P(d+1,k)$ that intersect $P(d,k)$ in a hyperplane of $P(d,k)$ tangent to O, (iii) the symbol ∞. Lines of $T(O)$ are of two types: (a) lines of $P(d+1,k)$ not in $P(d,k)$ that intersect $P(d,k)$

† See Benson (1970). ‡ See Thas and Payne (1976). § See Thas (1973*a*).

in a point of O, and (b) the points of O. Points of type (i) can only be on lines of type (a), and the incidence is determined by the usual incidence in $P(d+1, k)$. A point of type (ii) is in all lines of type (a) contained in it, and in the unique line of type (b) contained in it. The point ∞ is on all lines of type (b) and on no line of type (a).

The number of lines in $T(O)$ is easily seen to be $(k^{d-1}+1)(k^d+1)$.

For $d=2$, the number of tangent lines to a point of O in $P(d, k)$ is one. Each such line is on k planes not in the plane containing O. So there are $k(k+1)$ points of type (ii). Hence $v=(k+1)(k^2+1)$.

For $d=3$, a point of O is on one tangent plane in the 3-space on O, and each tangent plane is in k 3-spaces of $P(4, k)$ not counting the 3-space on O. So there are $k(k^2+1)$ points of type (ii). Thus $v=(k+1)(k^3+1)$.

It is not difficult to see that all lines have k points and that for $d=2$ we have $t=k$, and for $d=3$ we have $t=k^2$. We leave it as an exercise for the reader to check thoroughly that GQ1 holds.

We use the notation $T_d(O)$, $d=2$ or 3, to denote the above examples, which are due to Tits.

Now let O denote a *hyperoval* in $\Pi = P(2, k)$, k even; that is, a set of $k+2$ points no three collinear. Embed Π as a hyperplane in $S=P(3, k)$. Take as points the points of S not in Π and as lines, those lines of S not in Π which meet Π in a unique point of O. With incidence that of S, this gives a generalized quadrangle with parameters $(k-1, k+1)$. The dual has parameters $(k+1, k-1)$. These constructions are due to Ahrens and Szekeres (1969) and Hall (1971). There is a generalization due to Payne (1972) that is known to give genuinely different examples when k is a power of 2.

The third non-classical construction concerns *q-clans* which are special sets of 2×2 matrices over the field GF(q). To describe these, we need more information than we can give here, but we refer the enthusiastic reader to Payne (1972) and Kantor (1980). See also Payne (1990).

The fourth construction, known as *Roman generalized quadrangles,* have parameters (k, k^2) where $k=3^e$, $e>2$. These are due to Payne (1989) and again we leave the reader to seek out the construction.

6.3 Some combinatorial properties

Let $S=(P, L)$ be a generalized quadrangle. For $p\in P$, define $p^\perp$ (p 'perp') to be $\{x\in P \mid x\sim p\}$. (This notation was introduced in section 5.4. The more common notation for generalized quadrangles is $p^\perp = st(p)$, the *star* of p.)

For lines, in order to preserve duality, we define $\ell \sim h$ if ℓ and h intersect. Hence define $\ell^\perp = \{h\in L \mid h\sim \ell\}$. Note that $p\in p^\perp$ and $\ell \in \ell^\perp$.

The *trace* of a pair of distinct points p and q is defined by $\mathrm{tr}(p,q)=\{p,q\}^{\perp}$ $=\{x\in P|x\sim p$ and $x\sim q\}$. So, if $p\sim q$, $\mathrm{tr}(p,q)$ is the set of points of the line pq. More generally, for any subset X of P(or L), $X^{\perp}=\{x\in P(\text{or } L)|x\sim a$ for all $a\in X\}$. Notice, then, that $\ell^{\perp}$ can mean two different things, depending on whether ℓ is viewed as just an element of L, or as a set of points. Care must be taken to determine the context beforehand.

The results of this section will be proved for points, and then duality is invoked to claim the corresponding results for lines. Hence, we give the definitions below for points only, but the dual definitions will be assumed.

For distinct points p and q, the *span* of p and q is defined by $\mathrm{sp}(p,q)=\{\{p,q\}^{\perp}\}^{\perp}=\{x\in P|x\sim y$ for all $y\in\mathrm{tr}(p,q)\}$. Clearly, $p,q\in\mathrm{sp}(p,q)$ and, if $p\sim q$, $\mathrm{sp}(p,q)=pq$.

Consider the above concepts for a grid (figure 6.3.1). The set $p^{\perp}$ is the set of all points on the two lines on p, and so $|p^{\perp}|=s_1+s_2+1$, where s_1+1 and s_2+1 denote the numbers of points on 'horizontal' and 'vertical' lines respectively. For any line, $|\ell^{\perp}|=s_1+1$ or s_2+1. If $p\sim q$, then $|\mathrm{tr}(p,q)|=s_1+1$ or s_2+1. If $p\not\sim q, |\mathrm{tr}(p,q)|=2$. The span of adjacent points p and q is the line on p and q. If $p\not\sim q$, $\mathrm{sp}(p,q)=\{p,q\}$.

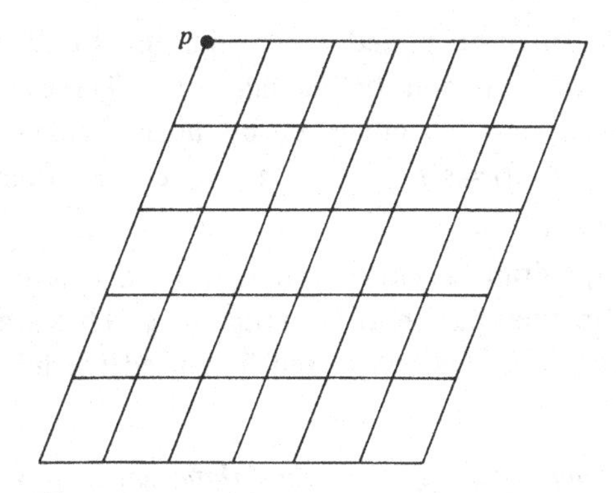

Figure 6.3.1.

Lemma 6.3.1. *If S is a generalized quadrangle with parameters s and t, then $|p^{\perp}|=st+s+1$ for any $p\in P$.*

Proof. The point p is on $t+1$ lines each having $s+1$ points. Excluding p, $(t+1)s$ points are therefore adjacent to p. □

Lemma 6.3.2. *If S is a generalized quadrangle with parameters s and t, then $|\mathrm{tr}(p,q)|=t+1$ for all $p\not\sim q$ in P.*

Proof. This is a restatement of lemma 6.1.2. □

Corollary. $|\mathrm{sp}(p,q)| \leq t+1$ for all $p \not\sim q$.

The pair (p,q) of distinct points is said to be *regular* if $|\mathrm{sp}(p,q)| = t+1$. Clearly, if $p \sim q$ and $s = t$, then (p,q) is regular.

The point p is *regular* if (p,q) is regular for all $q \neq p$.

The pair (p,q), $p \not\sim q$, is said to be *antiregular* if $|x^{\perp} \cap \mathrm{tr}(p,q)| \leq 2$ for all $x \in P \backslash \{p,q\}$.

The point p is *antiregular* if (p,q) is antiregular for all $q \neq p$.

The point p is *coregular* if each line on it is regular.

A *triad* is a triple of pairwise non-collinear points. For a fixed triad $T = (p,q,r)$, a *centre* of T is a point of $T^{\perp}$. We say T is *acentric, unicentric* or *centric* according as $|T^{\perp}| = 0, = 1$, or ≥ 1 respectively.

Lemma 6.3.3. *If S is a generalized quadrangle with parameters $s = t$ and if (p,q) is a regular pair, then (x,y) is a regular pair for all $x, y \in \mathrm{tr}(p,q)$.*

Proof. This is trivial if $p \sim q$. If $p \not\sim q$, it suffices to note that for any $u \in \mathrm{tr}(p,q)$, $u \sim w$ for all w in $\mathrm{sp}(p,q)$. Since $|\mathrm{sp}(p,q)| = |\mathrm{tr}(x,y)|$, we get $\mathrm{sp}(x,y) = \mathrm{tr}(p,q)$ and so $|\mathrm{sp}(x,y)| = s+1 = t+1$. □

Lemma 6.3.4. *Let S be a generalized quadrangle with parameters s and t. If $p \not\sim q$, then the pair (p,q) belongs to $s^2t - st - s + t$ triads.*

Proof. By lemma 6.3.1, each of p and q is adjacent to $st+s+1$ points, $t+1$ of which are common by lemma 6.3.2. Therefore the number of points not adjacent to either is, by using lemma 6.1.3, $(s+1)(st+1) - (1+2s+2st-t) = s^2t - st - s + t$, which is thus the number of triads on p and q. □

If S is a generalized quadrangle with parameters $s = t$, it is possible to show that, for any antiregular pair (p,q), each triad (p,q,r) has precisely zero or two centres. We prove this below, but first need the following lemma.

Lemma 6.3.5. *Let S be a generalized quadrangle with parameters $s = t$. Let p and q be non-adjacent points and set $\mathrm{tr}(p,q) = \{x_0, x_1, \ldots, x_s\}$. For $I \subseteq \{0, 1, \ldots, s\}$ let $n(I)$ be the number of points not collinear with p or q but collinear with precisely those x_i, $i \in I$. Then $\sum_I (|I|-1)(|I|-(s+1))n(I) = 0$.*

Proof. By lemma 6.3.4 there are $s^3 - s^2$ points not collinear with p or q. Hence $\sum_I n(I) = s^3 - s^2$. Of the $s+1$ lines through a fixed x_i, one goes through p and a second goes through q. Summing over all I on a fixed i, we have then $\sum_{i \in I} n(I) = s^2 - s$. This implies that (since $|\varnothing| = 0$)

$$\sum_I |I| n(I) = \sum_{i=0}^{s} \left(\sum_{i \in I} n(I) \right) = (s+1)(s^2 - s) = s^3 - s.$$

Finally, if $i \neq j$ there are $s-1$ points excluding p and q adjacent to both x_i and x_j by lemma 6.3.2. Thus, summing over the sets I containing a fixed pair i, j, $i \neq j$, yields

$$\sum_{\substack{i \neq j \\ i,j \in I}} n(I) = s - 1.$$

Summing over all pairs i, j, $i \neq j$ results $\left(\text{since} \binom{a}{b} = 0 \text{ by definition if } a < b\right)$ in

$$\sum_I \binom{|I|}{2} n(I) = \sum_{\substack{i \neq j \\ i,j \in I}} \left(\sum_{\substack{i \neq j \\ i,j \in I}} n(I) \right) = \binom{s+1}{2}(s-1).$$

So

$$\sum_I (|I|^2 - |I|) n(I) = s^3 - s.$$

Thus,

$$\begin{aligned} &\sum_I (|I| - 1)(|I| - (s+1)) n(I) \\ &\quad = \sum_I (|I|^2 - |I| - (s+1)|I| + s + 1) n(I) \\ &\quad = s^3 - s - (s+1)(s^3 - s) + (s+1)(s^3 - s^2) = 0. \quad \square \end{aligned}$$

Lemma 6.3.6. *If S is a generalized quadrangle with parameters $s = t$ and if S has an antiregular pair (p, q), then each triad (p, q, r) has zero or two centres.*

Proof. Since $p \nsim q$, (p, q) belongs to $s^3 - s^2$ triads by lemma 6.3.4.

Using the notation of lemma 6.3.5, (p, q) antiregular implies $n(I) = 0$ for $|I| > 2$. By lemma 6.3.5, then,

$$\begin{aligned} 0 &= \sum_I (|I| - 1)(|I| - (s+1)) n(I) = \sum_{|I| \leq 2} (|I| - 1)(|I| - (s+1)) n(I) \\ &= (s+1) n(\varnothing) + \sum_{|I| = 2} (2 - (s+1)) n(I). \end{aligned}$$

Hence

$$(s+1) n(\varnothing) = (s-1) \sum_{|I| = 2} n(I) = (s-1)\binom{s+1}{2}(s-1),$$

where $\binom{s+1}{2}$ is the number of ways of choosing I and, for each such choice, there are $s-1$ elements of $n(I)$ by lemma 6.3.2. Thus $n(\varnothing) = s(s-1)^2/2$ which represents the number of points r for which (p, q, r) is an acentric triad.

Moreover, the number of points r for which (p,q,r) has two centres is given by $\binom{s+1}{2}(s-1)$, where $\binom{s+1}{2}$ is the number of choices of pairs of points of I and $s-1$ is the number of ways r can be chosen adjacent to the two points of I chosen. Hence the sum of the number of points r for which (p, q, r) is an acentric triad and the number of points r for which (p,q,r) has two centres is $[s(s-1)^2/2]+[(s^3-s)/2]=s^3-s^2$, which accounts for all possible choices of r for which (p,q,r) is a triad. □

Corollary. *If (p,q,r) is a unicentric triad, then any two of p, q, r form a pair which is not antiregular.*

Lemma 6.3.7. *If S is a generalized quadrangle with parameters s and t, $1<s<t$, then no pair (p, q) is regular (and so no point is regular).*

Proof. Suppose (p,q) is regular. By counting points on lines joining points of $\mathrm{sp}(p,q)$ to points of $\mathrm{tr}(p,q)$ we obtain the inequality $v=(s+1)(st+1)\geq(t+1)(t+1)s$, or $ts(s+1)+1\geq ts(t+1)$. So $(s+1)+(1/st)\geq t+1$ implying, if $s>1$, $s>t$ which is a contradiction. □

Lemma 6.3.8. *Let S be a generalized quadrangle with parameters s and t. Then the pair (p,q) is regular with $s=1$ or $s\geq t$ if and only if each centric triad (p,q,r) has exactly 1 or $t+1$ centres.*

Proof. Suppose the pair (p, q) is regular. Further, suppose the triad (p, q, r) has at least two centres x and y. Then $|\mathrm{sp}(x, y)|=t+1$ since (p, q) is regular which implies that (p, q, r) has $t+1$ centres.

Conversely, suppose each centric triad (p, q, r) has either 1 or $t+1$ centres. If $t>1$, let x, y and z be any three points of $\mathrm{tr}(p,q)$. So (x,y,z) has $t+1$ centres and $|\mathrm{sp}(p,q)|=t+1$. If $t=1$, S is a grid and (p,q) is regular trivially. The fact that $s=1$ or $s\geq t$ follows from lemma 6.3.7. □

6.4 Generalized quadrangles with $s=t=3$

We assume throughout this section that S is a generalized quadrangle with parameters $s=t=3$. We shall classify all such quadrangles in terms of regularity conditions on the points and lines. All the results of this section are due to Payne (1975).

Notice that, by lemma 6.1.3, each quadrangle with $s=t=3$ has 40 points and 40 lines.

Lemma 6.4.1. *Every pair of points (lines) is regular or antiregular.*

Proof. Suppose there is a pair (p,q) which is neither regular nor antiregular.

Since (p,q) is not antiregular, there is a point x adjacent to more than two points of $\mathrm{tr}(p,q)$. Thus (p,q,x) forms a centric triad. By lemma 6.3.8, the triad cannot have four centres. Hence x is collinear with precisely three points of $\mathrm{tr}(p,q)$. Let $\mathrm{tr}(p,q) = \{x_0, x_1, x_2, x_3\}$ with $x \sim x_0, x_1, x_2$. The lines xx_0, xx_1 and xx_2 account for three distinct lines on x. The fourth must meet px_3 in a point and qx_3 in a point since none of the other lines on x do. Because of GQ1 the only possibility is that x meets px_3 and qx_3 in the same point, x_3, which is of course false. □

Lemma 6.4.2. *If* (p,q,r) *is any triad of points, then*

(i) *if* (p,q,r) *has one or four centres, any two of* p, q, r *form a regular pair.*
(ii) *if* (p,q,r) *has zero or two centres, any two of* p, q, r *form an antiregular pair.*
(iii) *no triad has exactly three centres.*
(iv) *either all pairs of* (p,q,r) *are regular or all pairs are antiregular.*

Proof. (iii) follows directly from the proof of lemma 6.4.1.

If (p,q,r) has one centre, then any two of p,q,r form a regular pair by the corollary to lemma 6.3.6 and by lemma 6.4.1.

If (p,q,r) has four centres, then clearly no pair is antiregular and the conclusion follows from lemma 6.4.1.

If (p,q,r) has zero or two centres then lemma 6.3.8 implies that each pair is antiregular.

Part (iv) follows from (i), (ii) and (iii). □

Lemma 6.4.3. *Either all pairs of points are regular, or all pairs of non-collinear points are antiregular. The dual holds for lines also.*

Proof. We suppose that not all pairs of non-collinear points are antiregular. So there is some regular pair (p,q), $p \not\sim q$. Let x and y be any two distinct points. We may assume $\{x,y\} \neq \{p,q\}$. If $x \sim y$ then clearly (x,y) is regular. If (x, p, q) is a triad then, by (iv) of lemma 6.4.2, (x, p) is regular. If (x,p,y) is then a triad, it follows by the same result that (x,y) is a regular pair. By lemma 6.3.3, we may assume that $\{x,y\} \not\subseteq \mathrm{tr}(p,q)$.

Now suppose that $x \not\sim y$ and that one of x and y, say x, is adjacent to both p and q, or that $x \sim p$ and $y \sim q$. We claim that in either case we can find a point w which is not adjacent to any of p, q, x, y. We show that w exists by counting points. There are 40 points in all. Thirteen points are adjacent to p including p itself. Four of these are already adjacent to q by GQ1, so we have, in addition, nine new points adjacent to q. If x is adjacent to both p and q, there are only two further lines on x, with six new points. Finally, y must be adjacent to a point on each line on p and q, adding at most ten new points. Thus we have accounted for at most 38 of the 40 points. If x is adjacent to p and y to q, the lines on x each meet a line on y and we have at most 14 additional points, accounting for 36 of the 40 in all. In either case, then, a w as claimed exists.

Now (p, q, w) is a triad and, by (iv) of lemma 6.4.2, (q, w) is a regular pair. Then (q, w, x) is a triad and by the same result (w, x) is a regular pair. Finally, (w, x, y) a triad implies (x, y) a regular pair. □

Lemma 6.4.4. *Any generalized quadrangle with parameters $s = t = 3$ has all pairs of non-collinear points antiregular and all pairs of lines regular, or is the dual of such a quadrangle.*

Proof. Suppose that all pairs of non-collinear points are antiregular. (This assumption is allowed by lemma 6.4.3.) Let ℓ_1 and ℓ_2 be any two lines. If $\ell_1 \cap \ell_2 \neq \emptyset$, then clearly $|\mathrm{sp}(\ell_1, \ell_2)| = t + 1$ and so (ℓ_1, ℓ_2) is a regular pair. Suppose $\ell_1 \cap \ell_2 = \emptyset$. Let $x \in \ell_1, y \in \ell_2$. Since $|\mathrm{tr}(x, y)| = 4$ (lemma 6.3.2), we can choose p and q, distinct points of $\mathrm{tr}(x, y) \backslash (\ell_1 \cup \ell_2)$. Now choose $z \in \mathrm{tr}(p, q)$, $z \neq x$ or y. Clearly, $z \notin \ell_1$ or ℓ_2. Then (x, y, z) is a centric triad and, since all pairs of non-collinear points are antiregular, (x, y, z) has precisely two centres (lemma 6.4.2), which are of course p and q. Let ℓ_3 be the line on z meeting ℓ_1. (See figure 6.4.1.) Let ℓ_4 be the line on x meeting ℓ_2. It is easy to check that $\ell_2 \neq xp$ or xq because of the fact that $p, q \notin \ell_2$. Let ℓ_5 be the line on z meeting ℓ_4. It is again an easy matter to check that ℓ_5 is not ℓ_3 or zp or zq. Notice also that $\ell_2 \cap \ell_4 \notin \ell_5$ as (x, y, z) has precisely two centres. Finally, let ℓ_6 be the line on y meeting ℓ_5. Then ℓ_6 is different from ℓ_2, yp and yq. As it is the fourth line on y and none of the other three lines

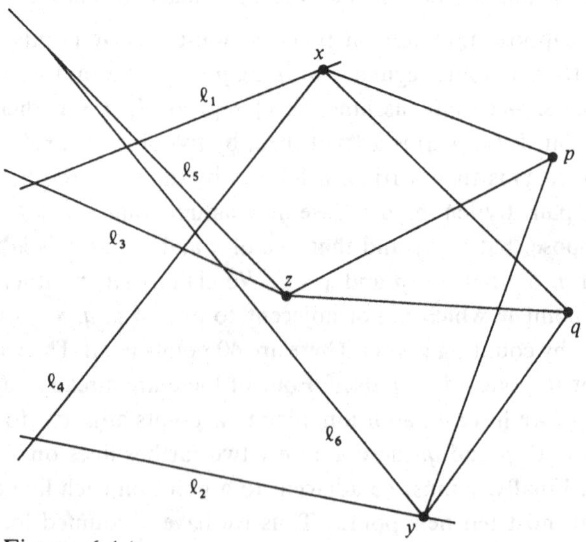

Figure 6.4.1.

on y meet ℓ_1, the line ℓ_6 meets ℓ_1. Moreover, ℓ_5 cannot meet ℓ_2 and, since pz and qz miss ℓ_2 also, the fourth line on z, ℓ_3, must meet ℓ_2. So (ℓ_1,ℓ_2,ℓ_5) is a triad of lines with centres ℓ_3, ℓ_4 and ℓ_6. In this case lemma 6.4.3 implies that in fact the triad has four centres and that (ℓ_1,ℓ_2) is regular as desired. □

If in a generalized quadrangle S with parameters $s = t > 1$, all pairs of points are regular (or, equivalently, all points are regular), it is known that S is isomorphic to $W(s)$ (Benson, 1970). Hence, for $s = t = 3$, our generalized quadrangle is isomorphic to $W(3)$ or to its dual, $Q(4, 3)$.

6.5 Subquadrangles

The near-linear space $S' = (P', L')$ is a *subquadrangle* of the generalized quadrangle $S = (P, L)$ if S' is a restriction of S and is itself a generalized quadrangle. S' is called *proper* if $S' \neq S$. Note that if S and S' have parameters s, t and s', t', respectively then $s = s'$ implies $L' \subseteq L$.

Example 6.5.1. Let S be a grid where the lines have either s_1 or s_2 points. Then S always contains a subquadrangle, itself a grid, for any line sizes s_1', s_2' where $1 \leq s_1' \leq s_1$ and $1 \leq s_2' \leq s_2$.

Dually a generalized quadrangle which is the dual of a grid always has subquadrangles with $s = s' = 1$ and $1 \leq t_1' \leq t_1$, $1 \leq t_2' \leq t_2$.

Example 6.5.2. Consider $Q(5, k)$ and set one of the variables equal to zero in the polynomial equation defining the quadric in 5-space. This results in a quadric defined in 4-space. Similarly, $Q(4, k)$ contains subquadrangles of the form $Q(3, k)$.

We move on now to some general results about parameters of subquadrangles of quadrangles. If S' is a proper subquadrangle of S it is not difficult to show that $v' < v$ and $b' < b$. We leave this as an exercise.

Theorem 6.5.1. *If the generalized quadrangle $S' = (P', L')$ with parameters s', t' is a subquadrangle of the generalized quadrangle $S = (P, L)$ with parameters s, t, then $s = s'$ or $s \geq s't'$, and dually $t = t'$ or $t \geq s't'$.*

Proof. Let $X = \{x \in P \backslash P' \mid x \text{ is not on any line of } L'\}$. We calculate $d = |X|$. The number of points in $P \backslash P'$ is $(s+1)(st+1) - (s'+1)(s't'+1)$. The number of points of $P \backslash P'$ on lines of L' is $(t'+1)(s't'+1)(s-s')$. So

$$d = (s+1)(st+1) - (s'+1)(s't'+1) - (t'+1)(s't'+1)(s-s') \geq 0.$$

If $t = t'$, then this reduces to $t(s-s')(s-s't) > 0$ and so $s \geq s't'$ or $s = s'$.

Suppose $t > t'$. We count the ordered pairs (x_i, z) for $x_i \in X$, $z \in P'$ with $z \sim x_i$. There are $(s'+1)$ $(s't'+1)$ choices for z and, for each of these,

$(t-t')s$ choices for x_i. Letting t_i be the number of points of P' collinear with x_i, we have

$$\sum_i t_i = (s'+1)(s't'+1)(t-t')s. \quad (1)$$

Now count the ordered triples (x_i, z, z') for $x_i \in X$ and for $z, z' \in P'$, with $x_i \sim z, z'$ and $z \nsim z'$. Again, there are $(s'+1)(s't'+1)$ choices for z, then $(s'+1)(s't'+1)-(s'(t'+1)+1) = s'^2t'$ choices for z'. Finally, there are $(t-t')$ choices for x_i. So

$$\sum_i t_i(t_i - 1) = (s'+1)(s't'+1)s'^2t'(t-t'). \quad (2)$$

Adding (1) and (2) gives $\sum_i t_i^2 = (s'+1)\,(s't'+1)\,(t-t')\,(s'^2t'+s)$.
By lemma 6.1.5, $d\sum_i t_i^2 \geq (\sum t_i)^2$ and so

$$(s'+1)(s't'+1)(t-t')(s-s')(s-s't')(st+s'^2t'^2) \geq 0.$$

Hence $t > t'$ implies $s = s'$ or $s - s't' \geq 0$. □

We are particularly interested in the case where a subquadrangle S' of a generalized quadrangle S is actually a subspace. This happens precisely when $s = s'$.

Lemma 6.5.2. *Let S' be a proper subquadrangle of S with parameters s, t' and s, t respectively. Then $t \geq s$, and $t = s$ implies $t' = 1$.*

Proof. We may suppose that $t' < t$.

If $t < s$, then $tt' < st' \leq t$ by theorem 6.5.1. Hence, $t' < 1$, a contradiction.

If $t = s$, then $t \geq st'$ by theorem 6.5.1, which implies $1 \geq t'$ and so $t' = 1$. □

Lemma 6.5.3. *Let S' be a proper subquadrangle of S with parameters s, t' and s, t respectively. If $s > 1$ then $t' \leq s$, and $t' = s$ implies $t = s^2$.*

Proof. If $t = 1$ then $t' \leq t$ implies $t' = 1 \leq s$ trivially. If $t > 1$ then theorems 6.5.1 and 6.1.6 imply $st' \leq t \leq s^2$ and so $t' \leq s$.

Suppose $s > 1$ and $t' = s$. So $1 < t' \leq t$ and theorem 6.1.6 implies $t \leq s^2$. But theorem 6.5.1 implies $t \geq s^2$ since $t' < t$. □

Lemma 6.5.4. *Let S' be a proper subquadrangle of S with parameters s, t' and s, t respectively. Then $s > 1$ and $t' > 1$ imply $s^{1/2} \leq t' \leq s$ and $s^{3/2} \leq t \leq s^2$.*

Proof. Theorem 6.1.6 applied to S' yields $s \leq t'^2$ and so $s^{1/2} \leq t'$. Lemma 6.5.3 gives $t' \leq s$.

Then $s^{1/2} \leq t'$ implies $s^{3/2} \leq st' \leq t$ by theorem 6.5.1. Finally, $t \leq s^2$ is from theorem 6.1.6. □

Lemma 6.5.5. *Let S' be a proper subquadrangle of S, and S'' a proper subquadrangle of S' where S'', S' and S have respectively parameters s, t''; s, t' and s, t; $s > 1$. Then $t'' = 1$, $t' = s$ and $t = s^2$.*

Proof. If $t'' > 1$, then applying lemma 6.5.4 twice yields $s^{3/2} \le t' \le s^2$ and $s^{1/2} \le t' \le s$, a contradiction. So $t'' = 1$. We may suppose then that $t > 1$.
By theorem 6.5.1, $t' \ge s$. By lemma 6.5.3, $t' \le s$. Hence $t' = s$. Again by lemma 6.5.3, $t = s^2$. □

The last lemma implies that any chain of proper subquadrangles $S \supsetneqq S' \supsetneqq S'' \supsetneqq \ldots$, in which $s > 1$ and the number of points on each line remains constant, has at most three elements. This is not true if $s = 1$ as can be seen by checking exercise 30 of section 6.8.

The next theorem gives us a method for determining when a subset of a generalized quadrangle is a subquadrangle with the same parameter s.

Theorem 6.5.6. *Let $S = (P, L)$ be a generalized quadrangle with parameters $s > 1$ and t. Let $S' = (P', L')$ where $P' \subseteq P$ and $L' \subseteq L$. Suppose*

(i) *$p, q \in P'$, $p \neq q$ and $p, q \in \ell \in L$ implies $\ell \in L'$;*
(ii) *for all $\ell' \in L'$, ℓ' contains $s+1$ elements of P'.*

Then one of the following holds.

(*a*) *All lines of L' are on a point p such that P' is the set of points of P collinear with p.*
(*b*) *$L' = \varnothing$.*
(*c*) *S' is a subquadrangle of S with parameters s and t'.*

Proof. It is easy to see that (a), (b) and (c) all satisfy (i) and (ii). We shall show that if S satisfies (i) and (ii) and is not of type (a) or (b), then it is of type (c). So suppose $P' \neq \varnothing \neq L'$.

To show that GQ1 holds for S', let p' be a point of P' not on the line $\ell' \in L'$. By GQ1 in S, and since $\ell \in L' \subseteq L$, there is a point $p \in P$ on ℓ' and a line ℓ of L on p and p'. By (ii), $p \in P'$. Then, by (i), $\ell \in L'$; and so GQ1 holds in S'.

Because of (ii) and theorem 6.1.1, to show that S' is a generalized quadrangle with parameters s and t', it suffices to show that each point of S' is on $t' + 1$ lines for some fixed number $t' \ge 1$.

Let p' be a point of P' on at least one line of L'. Since (b) is excluded, p' exists. Since (a) is excluded, there is a point q' not adjacent to p'. Let ℓ' be any line on p'. By GQ1 in S' there is a unique line of L' on q' meeting ℓ'. In fact, we can set up a 1–1 correspondence this way between lines on p' and lines on q'. Hence p' and q' are on the same number of lines, $t' + 1$ say, of S', where $t' \ge 0$.

If x' is any point of P' collinear with p' but not with q' then, as above, x' and q', and hence x' and p', are on $t'+1$ lines.

Finally, consider a point x' such that $x' \sim p', q'$. If $t'=0$ we claim that every line of L' is on x' and hence S' is of type (*a*) giving a contradiction. But it is easy to see that this happens, because, for any line ℓ' of L', $\ell' \neq p'x'$, $q'x'$, it is the case by GQ1 that p' (respectively q') is on a line meeting ℓ. But $t'=0$ implies $x' \in \ell'$ is the only possibility.

Consider $t' \geq 1$. Again, let $x' \sim p', q'$. Let ℓ' be a line on p' different from $x'p'$. By GQ1, q' is adjacent to a unique point of ℓ'. But $s>1$ implies there is a point $y' \in \ell'$ such that $y' \not\sim q'$. GQ1 also implies $y' \not\sim x'$. Now arguing for y' and q' as above for p' and q', we establish that y' and q' are on the same number of lines. Similarly, arguing for y' and x' yields the result that y' and x' are on the same number of lines. Hence p' and x' are on the same number of lines. □

The statement of theorem 6.5.6 is not valid if $s=1$. We leave this as an exercise for the reader.

In exercise 15 of section 5.8 we introduced the term ovoid. In projective 3-space $P(3,k)$ it is possible to show that ovoids have k^2+1 points (exercise 16 of section 5.8). We wish to accommodate this notion to generalized quadrangles.

Define an *ovaloid* of a generalized quadrangle S with parameters s and t to be a set of $st+1$ points of S no two of which are collinear.

It is not difficult to show that this is the largest possible set of this type in a generalized quadrangle. We leave this as an exercise.

Because of the duality available in generalized quadrangles, it makes sense to introduce also the next definition.

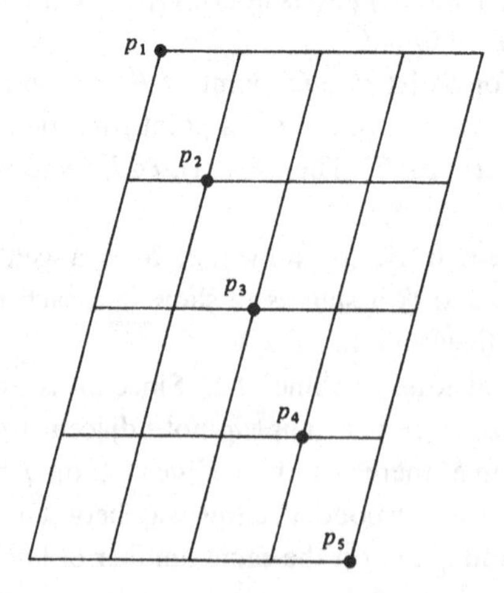

Figure 6.5.1.

A *spread* of a generalized quadrangle S with parameters s and t is a set of $st+1$ lines of S no two of which are concurrent.

Example 6.5.3. Let $t=1$. That is, S is a grid. Then $st+1=s+1$. It is not difficult to see that there are many ovaloids. In the quadrangle of figure 6.5.1, the set $\{p_1,p_2,\ldots,p_5\}$ is an ovaloid. There are $(s+1)!$ such sets. Moreover, the number of spreads is easily seen to be 2.

Example 6.5.4. Let $S=Q(4,k)\subseteq P(4,k)$. Here $st+1=k^2+1$. Let $P(3,k)$ be a hyperplane of $P(4,k)$ for which $P(3,k)\cap S=Q'$ is an elliptic quadric of $P(3,k)$. Hence Q' is a set of pairwise non-collinear points and by the results of section 5.6 has order k^2+1.

$Q(4,k)$ for k even is always self-dual, and hence has a spread. However, it is known that if k is odd there are no spreads. (See Thas (1972*b*).)

It is often difficult to decide whether or not a generalized quadrangle has ovaloids or spreads. There is a general result in this direction about ovaloids in subquadrangles, and we present it below.

Lemma 6.5.7. *Let $S'=(P',L')$ with parameters s, t' be a subquadrangle of the generalized quadrangle $S=(P,L)$ with parameter s, t, $t'<t$. Then S' contains an ovaloid.*

Proof. Let $p\in P\backslash P'$. Then p is not on any line of L'. There are $t+1$ lines on p in S and some of these may meet P'. However, such a line cannot meet P' in more than one point. Let $\ell_1,\ell_2,\ldots,\ell_r$ be the lines on p meeting P' in a unique point. It is easy to see that if $O=\{x_1,x_2,\ldots,x_r\}$ is the set of points in which these lines meet P', then no two points of this set are collinear. Moreover, each x_i is on $t'+1$ lines in S' and so, altogether, there are $r(t'+1)$ lines of S' incident with one of the points of O.

Let ℓ' be any line of L'. Let p' be the unique point of ℓ' such that pp' is a line of L. So p' is an element of O. Thus every line of L' meets a point of O. This implies $r(t'+1)=(t'+1)\ (st'+1)$, so $r=st'+1$ and O is an ovaloid of S'. □

6.6 Collineations of generalized quadrangles

Let f be a collineation of the generalized quadrangle $S=(P,L)$. Let

$$P_f=\{p\in P \mid f(p)=p\}.$$

Let $S_f=(P_f,L_f)$ be the restriction of S to P_f.

We are interested in this section in investigating the possibilities for the substructure S_f of S. We consider an example first of all.

Example 6.6.1. Let S be a grid with parameters $s\geq 3$, $t=1$. Label the points p_{ij}, $1 \le i, j \le s+1$, where, for each i, $\{p_{ij} | 1 \le j \le s+1\}$ is a line, and, for each j, $\{p_{ij} | 1 \le i \le s+1\}$ is a line. Let $f(p_{1j}) = p_{1j}$ and $f(p_{2j}) = p_{2j}$ for $1 \le j \le s+1$, while $f(p_{ij}) = p_{i+1j}$ for $3 \le i \le s$, $1 \le j \le s+1$, and $f(p_{s+1j}) = p_{3j}$ for $1 \le j \le s+1$. (See figure 6.6.1.)

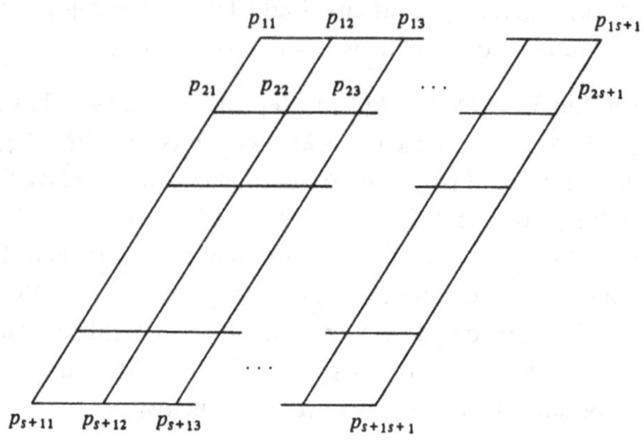

Figure 6.6.1.

The lines $\{p_{ij} | 1 \le j \le s+1\}$ for $i = 1$ and 2 are fixed pointwise. The other 'horizontal' lines are rotated. The 'vertical' lines are fixed by f but not pointwise. The restriction S_f of S to P_f is the $2 \times (s+1)$ grid given by the first two horizontal lines.

Theorem 6.6.1. *Let f be a collineation of the generalized quadrangle $S = (P, L)$. Let S_f be the restriction of S to the fixed points of f. Then one of the following holds.*

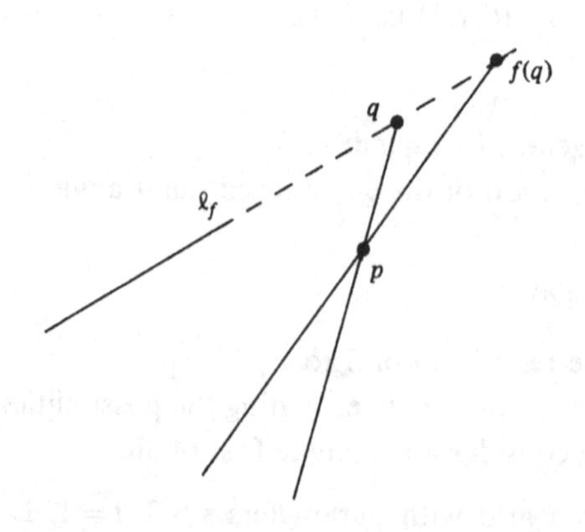

Figure 6.6.2.

(a) S_f *is a generalized quadrangle.*

(b) S_f *is a set of points no two collinear.*

(c) S_f *is a set of lines no two intersecting.*

(d) S_f *is a set of lines intersecting in a common point.*

Proof. We may suppose that ℓ_f is a line of S_f and p a point of S_f not on ℓ_f. By GQ1 there is a line of S on p meeting a point q of ℓ, where ℓ_f is the restriction of the line ℓ to f. If $f(q) \neq q$, then since $f(q) \in \ell$ and $f(p) = p$, we have a second line on p meeting ℓ, a contradiction. So $q = f(q)$. Hence $q \in \ell_f$ and therefore the sets of points $(pq)_f$ is a line in S_f. So GQ1 holds in S_f. □

A collineation f is said to be a *whorl about a point* p of S provided f fixes each line of S incident with p. In the above example 6.6.1, f is a whorl about each of p_{ij}, $i = 1, 2$ and $1 \leq j \leq s+1$.

Let f be a whorl about p. If f is the identity function or if $f(x) \neq x$ for all points $x \not\sim p$, then f is called an *elation about* p. (There is *some* similarity here between this definition of elation and the one used for projective planes.)

Furthermore, if f is a collineation which fixes each point of $p^\perp$, then f is called a *symmetry about* p.

The function f of example 6.6.1 is neither an elation nor a symmetry about any one of its points. It is possible to show that as soon as $s > 1$ any symmetry about p is automatically an elation about p. We shall give the proof below, but first state a result whose verification is left to the reader.

Lemma 6.6.2. *The set of symmetries about a point p forms a group.*

Lemma 6.6.3. *For $s > 1$ any symmetry about a point p is an elation about p.*

Proof. Let q be a point not adjacent to p such that the symmetry f fixes q. We show first of all that f fixes each point of $q^\perp$. Let x be such a point, $x \neq q$, $x \not\sim p$. Consider a line ℓ on p not meeting qx. By GQ1 there is a point x' on it such that $x \sim x'$. Now $f(x) \in f(qx) = qf(x)$ and $f(x) \in f(xx') = f(x)x'$. So, if $f(x) \neq x$, we contradict GQ1.

Now consider any point $x \notin \mathrm{sp}(p, q)$. By GQ1, x is adjacent to some point q' of $q^\perp$, $q' \not\sim p$. Arguing as above with q' replacing q this time, we see that $f(x) = x$.

Finally, suppose $x \in \mathrm{sp}(p, q)$. Let $y \in \mathrm{tr}(p, q)$. Since $s > 1$, there is a point $x' \in xy \setminus \{x, y\}$ and clearly $x' \notin \mathrm{sp}(p, q)$. Hence $f(x') = x'$. Moreover, $x' \not\sim p$ and so, again arguing as above using x' as q, $f(x) = x$. □

To complete the connection between types of collineations in projective planes and types of collineations in generalized quadrangles, we make the following definition.

Let f be a whorl about the point p for which there is a point $q \not\sim p$ with $f(q)=q$. If f is the identity or if q is unique, then f is a *homology* about p. The reader is urged to compare this definition with that for projective planes. We point out that the function of example 6.6.1 is not a homology.

6.7 A brief history of generalized quadrangles

Generalized quadrangles arose in two separate areas and as special cases of two different theories. They were first introduced by Tits (1959) as a special case of structures called 'generalized n-gons'. The definition of generalized n-gon, though quite uncomplicated, needs to be well motivated and so, rather than give it here, we shall permit the interested reader to do some research privately by referring to Tits (1959) or Dembowski (1968). Suffice it to say that it arose in connection with some problems in the theory of groups.

Generalized quadrangles arose again in work by Bose (1963). Bose dealt with systems he called partial geometries, and we shall be meeting these in chapter 7. Very generally, the original motivation for the theory from which they arose was from problems in statistics.

We also mentioned in chapter 5 that it was possible to view generalized quadrangles as those polar spaces with rank 2.

The major work on these structures has taken place in the last ten years and has been led for the most part by Thas, Payne and Tallini. We refer the reader to the many articles listed under these names in the bibliography.

As mentioned in section 6.2, all of the classical examples are due to Tits. Ahrens and Szekeres (1969) discovered a new class of quadrangles with parameters $k-1$, $k+1$, k a prime power. M. Hall (1971) gave this independently for k even. Payne (1972, 1974) generalized this to produce more examples. The newest class of examples is due to Kantor (1980). This was discovered in 1979 and the parameters are k^2, k where $k \equiv 2$ (mod 3).

One of the key problems is to give a classification of generalized quadrangles. Most of the known classifications of the classical examples are due to Thas (1975*b*, 1977*b*, 1978*b*). Payne has been the principal investigator in an attempt to study and to find classifications of generalized quadrangles in terms of their collineation groups. The reader is referred to Payne (1982, 1990) for a survey of work here. Much work has also been done however by M. Walker (1977), Thas (1979*a*) and Tits (1974).

6.8 Exercises

1. Show that not all polar spaces are generalized quadrangles.
2. Show that the generalized quadrangle of figure 6.1.3 is isomorphic to Sylvester's geometry.
3. Prove that the generalized quadrangles are precisely the polar spaces of rank 2.
4. Calculate the lines in the generalized quadrangle of example 6.2.1.
5. Construct $W(2)$ and draw a diagram.
6. Check that, for $T(O)$, GQ1 holds.
7. Prove that if $p \not\sim q$ then $x \not\sim y$ for all x and y in $\mathrm{tr}(p,q)$.
8. Compute the orders of $p^\perp$, $\mathrm{tr}(p,q)$ and $\mathrm{sp}(p,q)$ for distinct points p and q in a dual grid.
9. Compute the orders as in exercise 8 for a non-singular quadric of rank 2 in $P(4,k)$.
10. Show that if $t = s^2$, then the t_i's defined in theorem 6.1.6 are all equal to s, and each triad has $s+1$ centres.
11. In an arbitrary grid, determine the regular pairs of points and regular points, the antiregular pairs of points and antiregular points, and the coregular pairs of points.
12. Show that it is possible for a point to be both regular and antiregular.
13. Show that non-adjacent points p and q are regular if $|x^\perp \cap \mathrm{tr}(p,q)| \geq 2$ implies $\mathrm{tr}(x,y) \subseteq x^\perp$.
14. Let p be any regular point of $S = (P, L)$ with parameters s and t. Let ℓ_1 and ℓ_2 be lines on p. Define $S^* = (P^*, L^*)$ by $P^* = \{x \in P \mid x \not\sim p\}$, $L^* = \{l \in L \mid p \notin l\} \cup \{\mathrm{tr}(x_1, x_2) \setminus \{x\} \mid x_i \in \ell_i \setminus \{x\}, i = 1, 2\}$. Prove that S^* is a generalized quadrangle with parameters $s-1$, $t+1$.
15. Let p be a regular point of the generalized quadrangle $S = (P, L)$ with parameters s and t. Define $S' = (P', L')$ by $P' = p^\perp$, $L' = \{\mathrm{sp}(x,y) \mid x \neq y, x \sim p \sim y\}$. Show that S' is a projective plane.
16. Let $S = (P, L)$ be a generalized quadrangle with parameters s, t, in which all pairs of non-adjacent points are regular (equivalently, all points are regular). Define a new structure $\bar{S} = (\bar{P}, \bar{L})$ by $\bar{P} = P$ and $\bar{L} = L \cup \{\mathrm{sp}(x,y) \mid x \not\sim y\}$. Show that $\bar{S}$ is a projective 3-space.
17. Show that each line and point of $Q(3,k)$ is regular.
18. Show that each line of $Q(4,k)$ is regular.
19. If k is even, show that each point of $Q(4,k)$ is regular; while, if k is odd, show that each point is antiregular.

20. Count the number of triads in (*a*) a grid, (*b*) a dual grid.

21. Define the *trace* of triad (p, q, r) to be $\mathrm{tr}(p, q, r) = \{x | x \sim p, q \text{ and } r\}$. Show that $|\mathrm{tr}(p, q, r)| = s + 1$ if $s^2 = t$, $s > 1$.

22. Define the *span* of a triad (p, q, r) to be $\mathrm{sp}(p, q, r) = \{x | x \sim y \text{ for all } y \in \mathrm{tr}(p, q, r)\}$. Show that $|\mathrm{sp}(p, q, r)| \leq s + 1$.

23. Show that in any generalized quadrangle with parameters $s > 1$, t, it is always possible to find centric triads of points (of lines).

24. Show that in a generalized quadrangle with parameters s, t, $1 < s < t$, no pair of non-collinear points is antiregular.

25. Prove that for a generalized quadrangle with parameters s,t the pair $\{p,q\} p \not\sim q$ is regular if and only if each triad (p,q,r) is centric.

26. Prove directly (without using duality) that $b = (t + 1)(st + 1)$ in lemma 6.1.3.

27. Prove directly the dual of lemma 6.3.4.

28. Prove directly the dual of lemma 6.3.7.

29. If S' is a proper subquadrangle of S show that $v' < v$ and $b' < b$.

30. Let S be a dual grid. Show that S has a proper subquadrangle S' for all $1 \leq t' < t$.

31. Let S' with parameters s,t' be a proper subquadrangle of S with parameters s,t. Show that $s > 1$, $t' > 1$ and $t = s^{3/2}$ imply $t' = s^{1/2}$.

32. In the proof of theorem 6.5.1, show that $s = s'$ implies $\bar{t} = st' + 1$. ($\bar{t}$ was defined in lemma 6.1.5.)

33. Show that the conclusion of theorem 6.5.6 does not hold if $s = 1$.

34. Show that if X is a set of points of a generalized quadrangle with parameters s, t and no two points of X are collinear then $|X| \leq st + 1$.

35. Find the ovaloids and spreads in the example of figure 6.1.3.

36. Show that a grid with $s + 1$ points per line has $(s + 1)!$ ovaloids.

37. Show that there are generalized quadrangles $S = (P, L)$ and $S' = (P', L')$ with parameters s, t and s', t' respectively such that $P' \subseteq P$ but $L' \nsubseteq L$.

38. Let $S = (P, L)$ and $S' = (P', L')$ be generalized quadrangles with parameters s, t and s,t' respectively. Suppose $P' \subseteq P$ and $L' \subseteq L$. Show that, for any $\ell \in L$, there are only three possibilities: there is no point of P' on ℓ; there is a unique point of P' on ℓ; $\ell \in L'$.

39. Under the conditions of exercise 38, show dually that for any point $p \in P$ there are only three possibilities: there is no line of L' on p; there is a unique line of L' on p; $p \in P'$.

40. Prove that the set of symmetries about a point p of a generalized quadrangle S forms a group.

41. Prove that the set of homologies about a point p of a generalized quadrangle does not necessarily form a group.

42. If f is a collineation of the generalized quadrangle S with parameters $s > 1$, t such that $f(x) = x$ for all points x adjacent to p or q, where f is a whorl about p and about q, $p \not\sim q$, show that f is the identity.

43. Show that the statement of exercise 42 is not necessarily true if $s = 1$.

44. If f is a homology about p, and q is the unique element of $P \backslash p^{\perp}$ such that $f(q) = q$, then show that f is a homology about p.

45. Prove that Hermitian polarities exist in $P(d, k)$ if and only if k is a square.

46. Show that the number of m-dimensional subspaces of $P(d, q^2)$ in the polar space arising from a Hermitian polarity is as in section 6.2.

47. Show that a polarity is symplectic if and only if every point is absolute.

48. Show that $P(d, k)$ admits symplectic polarities if and only if d is odd.

49. Prove that the number of m-dimensional subspaces of $P(d, k)$ in the polar space arising from a symplectic polarity is as in section 6.2.

50. Define a *correlation* of a generalized quadrangle with $v = b$ to be a 1–1 map of the set of points and lines to itself, such that points map to lines and lines to points. (Compare with section 5.5.) A *polarity* is a correlation of order 2. Show that the generalized quadrangle of figure 6.1.3 is *self-polar*: that is, it is always isomorphic to any polarity of itself.

51. An *absolute point* (*line*) of a polarity σ (see exercise 50) is a point p (line ℓ) satisfying $p \in \sigma(p)$ ($\ell \ni \sigma(\ell)$). Show that the set of absolute lines forms a spread. (Hint: show first that a line cannot contain two absolute points. Then show that if ℓ is not absolute, the point $\sigma(\ell)$ is on a line which meets ℓ in an absolute point.)

7

Partial geometries

'What's the good of Mercators, North Poles
and Equators
Tropics, Zones, and Meridian Lines?'
So the Bellman would cry: and the crew
would reply
'They are merely conventional signs!'

Lewis Carroll *The Bellman's Speech*

In this, the final chapter, we consider a generalization of the concept of generalized quadrangle. Generalized quadrangles were first introduced by Tits (1959). The generalization, partial geometry, first appeared independently of Tits' work in a paper by Bose (1963).

7.1 The definition

A *partial geometry* is a finite near-linear space $S = (P, L)$ such that

PG1 for each point p and line ℓ, $p \notin \ell$ implies $c(p, \ell) = \alpha$,

PG2 each line has $s + 1$ points,

PG3 each point is on $t + 1$ lines,

where α, s and t are fixed positive integers.

Note that $\varnothing$ is trivially a partial geometry. If S is a partial geometry which is not $\varnothing$, we say S has *parameters* α, s and t.

We note also that, as for generalized quadrangles, the dual of a partial geometry is again a partial geometry.

Clearly, if $\alpha = 1$ and GQ2 is satisfied, then S is a generalized quadrangle. It is not difficult to show that if $S \neq \varnothing$, and $\alpha = 1$, then GQ2 *is* satisfied, and we leave this as an exercise.

From the definition of α, s and t, we see that $\alpha \leq s + 1$ and $\alpha \leq t + 1$. So $\alpha \leq \min\{s, t\} + 1$. The family of partial geometries is then conventionally divided into four (non-disjoint) subfamilies as follows:

(*a*) The partial geometries with $\alpha = s + 1$, or dually $\alpha = t + 1$. If $\alpha = s + 1$, S is a linear space. Projective and affine spaces are examples of this type of partial geometry. Note that if $\alpha = s + 1$ then $t \geq s$, whereas $\alpha = t + 1$ implies $s \geq t$.

(*b*) The partial geometries with $\alpha = s$, or dually $\alpha = t$. In case $\alpha = s$, such structures are included in the class of *copolar spaces* first introduced by J. I. Hall (1982). If $\alpha = t$, it is not difficult to show that the relation 'parallel' on the set L is an equivalence relation. (We leave this as an exercise.) In this case the partial geometry is called a *net* (Bose, 1963).

(*c*) The partial geometries with $\alpha = 1$. As mentioned earlier, these are the generalized quadrangles with parameters s and t.

(*d*) The partial geometries with $1 < \alpha < \min\{s, t\}$. These are known as the *proper partial geometries.*

Lemma 7.1.1. *In a partial geometry with parameters* α, s *and* t, $v = (s+1)(st+\alpha)/\alpha$ *and* $b = (t+1)(st+\alpha)/\alpha$ (Thas, 1977*a*).

Proof. (This proof generalizes that of lemma 6.1.3.) Fix a line ℓ. There are $v-(s+1)$ points not on ℓ. Each point not on ℓ is on α lines meeting ℓ. Hence, an alternative way to count all the points not on ℓ is to count all the points on lines meeting ℓ and then divide by α. There are $(s+1)t$ such lines and each has s points not on ℓ. Hence $v-(s+1) = st(s+1)/\alpha$ or $v = (s+1)(st+\alpha)/\alpha$. Dually, $b = (t+1)(st+\alpha)/\alpha$. □

Corollary. *In a partial geometry with parameters* α, s *and* t, $\alpha | st(s+1)$ *and* $\alpha | st(t+1)$.

Example 7.1.1. The case $\alpha = 1$ was thoroughly considered in chapter 6. So consider $\alpha = 2$. If $s = t = 1$, lemma 7.1.1 implies $v = b = 3$, and so S is a triangle. If $s = 1$ and $t = 2$, then $v = 4$ and $b = 6$. This is the affine plane of order 2. Hence if $s = 2$ and $t = 1$, S is the dual affine plane of order 2, with $v = 6$ and $b = 4$, shown in figure 7.1.1.

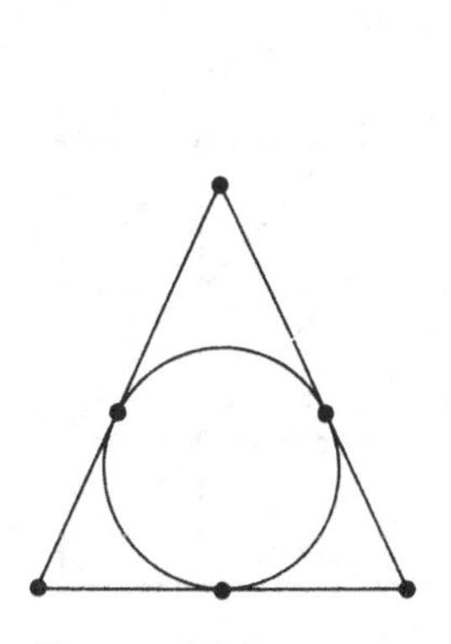

Figure 7.1.1.

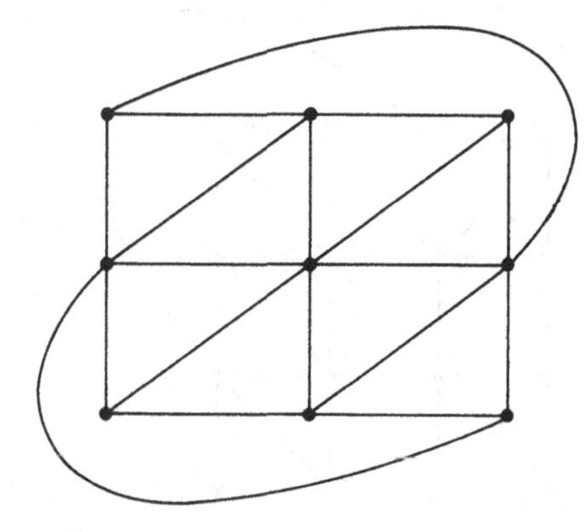

Figure 7.1.2.

Example 7.1.2. If $\alpha = s = t = 2$, then $v = b = 9$. Figure 7.1.2 represents such a partial geometry. It is the affine plane of order 3 minus one of its parallel classes. We leave it to the reader to show that this partial geometry is unique.

Example 7.1.3. If $\alpha = t = 2$ and $s = 3$, then $v = 16$, $b = 12$. An example here is given by the affine plane of order 4 minus eight of its lines comprising two parallel classes. See figure 7.1.3(*a*). There is exactly one other example with the same parameters, due to Shrikhande (1959). It is shown in figure 7.1.3(*b*). Both examples can be obtained from *Latin squares* (exercise 53 of section 4.8) of order 4, which in turn correspond to the multiplication tables of the two unique groups of order 4: $\{e, a, a^2, a^3\}$ with $a^4 = e$; and $\{e, a, b, ab\}$ with $a^2 = b^2 = e$ and $ab = ba$.

The Shrikhande example cannot be embedded in an affine plane of order 4.

Example 7.1.4. We generalize example 7.1.2 and the example of figure 7.1.3(*b*) in example 7.1.3. Let Π be any finite projective plane. Fix a line ℓ. Let $W \subseteq \ell, |W| \geq 2$. Consider the substructure on the points of $\Pi \backslash \ell$ where lines are lines of Π on some point of W, restricted to $\Pi \backslash \ell$. Clearly, if Π has order k, then each line of the substructure has k points. Fix a point p not on a line h. There is a unique line of the substructure on p missing h. There are $|W| - 1 \geq 1$ lines on p meeting h. Hence $\alpha = |W| - 1 = t$ and the substructure is a partial geometry. Such a partial geometry is actually a *net*. For the definition, see exercise 44 of section 7.7.

Example 7.1.5. Let P be the set of points of a projective space $P(d, k)$, which are not contained in a fixed subspace $P(d-2, k), d \geq 3$. Let L be the set of lines of $P(d, k)$ which do not have a point in common with

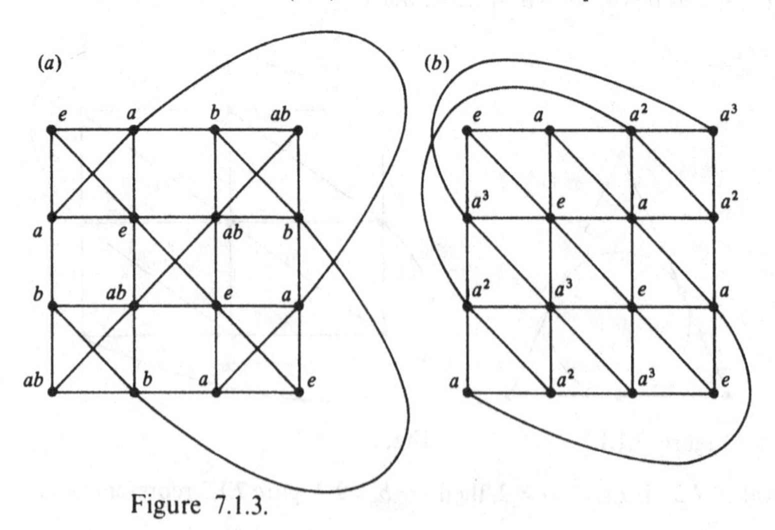

Figure 7.1.3.

$P(d-2,k)$. Then $S=(P,L)$ is a partial geometry with $\alpha=s=k$, $t=k^{d-1}-1$. (We leave the details to the reader.) We shall refer to this example as H_k^d (Thas and DeClerck, 1977).

The result of lemma 6.1.4 for the generalized quadrangle case can be immediately generalized to the partial geometry case. The proof is identical except for the fact that α replaces 1. We state the result and leave the proof to the reader.

Lemma 7.1.2. *If S is a partial geometry with parameters α, s and t, then* $\alpha(s+t+1-\alpha)|st(s+1)(t+1)$.

7.2 A method of constructing proper partial geometries

The theory of partial geometries first appeared in 1963 in a paper by Bose. (We mentioned this in section 6.7.) But until 1973 no examples of proper partial geometries were known. Then Thas published a method of constructing them using arcs in projective planes. (See Thas (1973*a*, *b*; 1974*a*).) In order to describe his results here, we need to define and give some information about arcs.

An *$\{n;\, d\}$-arc K*, $nd>0$ in a finite projective plane of order k is a set of n points such that d is the greatest number of collinear points in the set. Fix a point p of K. Then each line on p has at most d points of K and so $n\leq(d-1)(k+1)+1$. If $n=(d-1)(k+1)+1=dk-k+d$, then K meets each line of P either in no points or in d points. In this case we call K a *maximal $\{n;d\}$-arc*.

Lemma 7.2.1. *Let K be a $\{dk-k+d;d\}$-arc of a projective plane Π of order k with $1<d<k$. Let P be the set of points of Π not contained in K. Let L be the set of lines of Π incident with d points of K, each restricted to P. Then $S=(P,\ L)$ is a partial geometry with parameters $s=k-d$, $t=k-k/d$, $\alpha=k-k/d-d+1$.*

Proof. It is clear that $s=k-d$. To calculate t, fix a point p not on K. The number of lines on this point meeting K in d points is $(dk-k+d)/d=k-k/d+1=t+1$. Hence, the number of lines on p which meet points of K on a fixed line l not on p, is $k-k/d+1-d$. □

Corollary. *A necessary condition for the existence of a $\{dk-k+d;\ d\}$-arc in a projective plane of order k, is $d|k$.*

The condition presented in the corollary above is not a sufficient one. Cossu (1960) has shown that there is no $\{21;\ 3\}$-arc in the Desarguesian plane $P(2, 9)$. In general, Thas (1975*a*) has proved that there is no $\{2q+3;\ 3\}$-arc in a plane $P(2,\ k)$, $k=3^h$, $h>1$.

Lemma 7.2.2. *Let K be a $\{kd-k+d;d\}$-arc, $1<d<k$, in the projective*

plane $P(2,k)$. Embed $P(2,k)$ as a plane in $P(3,k)$. Let P be the set of points of $P(3,k)$ not in $P(2,k)$. Let L be the set of lines of $P(3,k)$ which are not contained in $P(2,k)$ and which meet K. Then $S=(P,L)$ is a partial geometry with parameters $\alpha=d-1$, $s=k-1$, $t=(k+1)(d-1)$.

Proof. It is clear that $s=k-1$, and also that $t=(k+1)(d-1)$. To compute α, fix a point $p\in P$ and a line $\ell\in L$, $p\notin L$. Then $\langle p,\ell\rangle$ is a plane which meets $P(2,k)$ in a line h(lemma 3.9.8). Since l meets K, h intersects K in d points. Also, as all lines of L on p meeting ℓ also meet K and are in $\langle p,\ell\rangle$, this means that $\alpha=d-1$. □

Suppose K is a $\{kd-k+d;d\}$-arc of a projective plane Π of order k where $d\le k$. Consider $\{\ell\in\Pi\,|\,\ell\cap K=\varnothing\}$. Each point of K is on $k+1$ lines each with d points in K. So the number of lines meeting K is $(k+1)(kd-k+d)/d$. Hence, the order of the above set is $k^2+k+1-(k+1)(kd-k+d)/d=k(k-d+1)/d$. For any point $p\notin K$, $(kd-k+d)/d$ lines on p meet K, and so $k+1-(kd-k+d)/d=k/d$ do not meet K. Now consider the dual plane Π^* of Π. We have just proved that the above set of lines having no intersection with K forms a $\{k(k-d+1)/d;k/d\}$-arc in Π^*. We summarize this as follows.

Lemma 7.2.3. *If K is a $\{kd-k+d;\ d\}$-arc of a projective plane Π of order k, $1\le d\le k$, then the dual plane contains a $\{k(k-d+1)/d;\ k/d\}$-arc.*

It is still not known when partial geometries contain maximal arcs. Denniston (1969) showed that $\{2^{m+h}-2^h+2^m;2^m\}$-arcs exist in $P(2,2^h)$ for all $0\le m\le h$ and so there exist partial geometries with parameters

$$\alpha=(2^m-1)(2^{h-m}-1), s=2^h-2^m, t=2^h-2^{h-m} \text{ and } \alpha=2^{m-1},$$
$$s=2^h-1, t=2^{h+m}-2^h+2^m-1, 0<m<h.^\dagger$$

More recently De Clerck, Dye and Thas (1980) discovered a new infinite class of partial geometries associated with hyperbolic quadrics in $P(4d-1,2)$. The parameters are $\alpha=2^{2n-2}, s=2^{2d}-1,\ t=2^{2d-1}$‡. To discuss their construction, we need more information about quadrics than has been covered in this text. The paper concerned would make a good seminar topic, however, for someone willing to learn more about quadrics.

7.3 Strongly regular graphs

It is often very convenient to switch to the notation and terminology of graph theory when dealing with near-linear spaces with a good deal of

† Thas gave other constructions for some Desarguesian planes and non-Desarguesian planes of order 2^m.

‡ Thas has given a construction in $P(4d-1,3)$. Constructions with parameters $\alpha=2$, $s=t=5, v=81$ due to van Lint and Schrijver (1980), and with parameters $\alpha=2$, $s=4, t=17, v=175$ due to Haemers (1981), also exist.

structure. This is especially true of partial geometries which can be re-interpreted as graphs with very nice properties. Hence we have the following definition.

A *strongly regular graph* is a graph $G = (V, E)$ of points (vertices) V and lines (edges) E such that there are non-negative integers k, λ and μ satisfying

(i) each point is adjacent to k other points,
(ii) any two adjacent points are mutually adjacent to λ other points,
(iii) any two non-adjacent points are mutually adjacent to μ points.

We call k, λ and μ the *parameters* of G.

Lemma 7.3.1. *A strongly regular graph with v points and having parameters k, λ and μ satisfies $(v - k - 1)\mu = k(k - 1 - \lambda)$, or equivalently $v = (k(k - 1 - \lambda + \mu) + \mu)/\mu$.*

Proof. Let $\Gamma(p)$ be the set of points adjacent to a fixed point p, excluding p itself, and let $\Delta(p)$ be the set of points not adjacent to p. We count in two different ways the number of lines which have one point in $\Delta(p)$ and one point in $\Gamma(p)$. (Remember that lines have only two points here!) There are $v - (k + 1)$ points in $\Delta(p)$ and each of these is adjacent to μ points of $\Gamma(p)$, yielding a total of $(v - k - 1)\mu$ lines. However, any point x in $\Gamma(p)$ is adjacent to λ other points of $\Gamma(p)$. The other $k - 1 - \lambda$ points adjacent to x are all in $\Delta(p)$. Since this occurs for each of the k points of $\Gamma(p)$, we have a total of $k(k - 1 - \lambda)$ lines. Hence the first equality in the statement of the lemma is true. □

Example 7.3.1. Let $k = 2$, $\lambda = 0$, $\mu = 1$, and so $v = 5$ by lemma 7.3.1. This graph is shown in figure 7.3.1. It is called the *pentagon graph.*

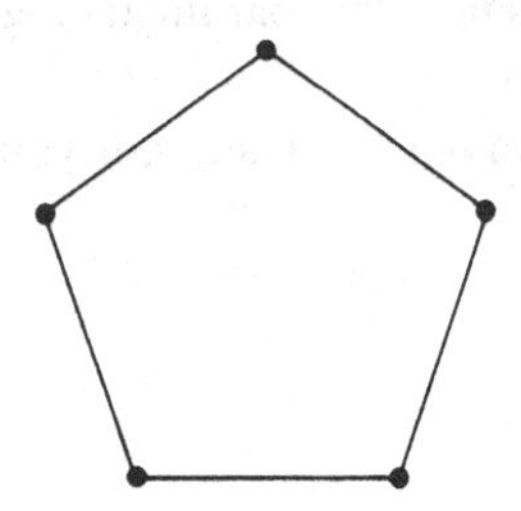

Figure 7.3.1.

See exercise 21 of section 7.7 for a third example of a strongly regular graph.

Example 7.3.2. Let $k = 3$, $\lambda = 0$, $\mu = 1$, and so $v = 10$ by lemma 7.3.1. This graph is unique and is called the *Petersen graph* (figure 7.3.2).

Lemma 7.3.2. *A partial geometry with parameters α, s, t forms a strongly* *regular graph with* $v = (s+1)(st+\alpha)/\alpha$ *and with parameters* $k = s(t+1)$, $\lambda = t(\alpha-1)+s-1$, $\mu = \alpha(t+1)$ *where adjacency in the graph is given by* $p \sim q$ *if and only if* $p \neq q$ *and* p *and* q *are collinear.*

Proof. The value for v follows from lemma 7.1.1. Then k can be evaluated by counting the points on the $t+1$ lines on any fixed point p. For two adjacent points p and q, consider the t lines on p excluding the line pq and count the points on those lines which are also adjacent to q. (There are α such points per line.) If p and q are non-adjacent, again count the points on the $t+1$ lines on p which are also adjacent to q. □

As we saw in chapter 1 it is possible to construct a graph from any near-linear space using the lines of the space. This was called the line graph of the space and, as we see in the next lemma, the line graph of a partial geometry forms a strongly regular graph. We shall sometimes call the graph of lemma 7.3.2 the *point graph* of the partial geometry.

Lemma 7.3.3. *A partial geometry S with parameters* α, s, t *has a strongly regular line graph with* $v = (t+1)(st+\alpha)/\alpha$ *and with parameters* $k = t(s+1)$, $\lambda = s(\alpha-1)+t-1$, $\mu = \alpha(s+1)$.

Proof. The value for v follows from lemma 7.1.1. Since each point of a fixed line ℓ is on t lines other than ℓ, $k = t(s+1)$. For any two intersecting lines ℓ and $\hbar$, there are $t-1$ other lines on the point of intersection, and $s(\alpha-1)$ lines meeting both ℓ and $\hbar$ in distinct points. If ℓ and $\hbar$ do not intersect, then there are $(s+1)\alpha$ lines meeting both. □

An alternative way of proving lemma 7.3.3 would have been to point out that the line graph of the partial geometry S is the same as the point graph of the dual partial geometry. Hence the parameters given in lemma 7.3.3 are the dual of those in lemma 7.3.2.

The converse of lemma 7.3.2 is not always true. That is, if you have a

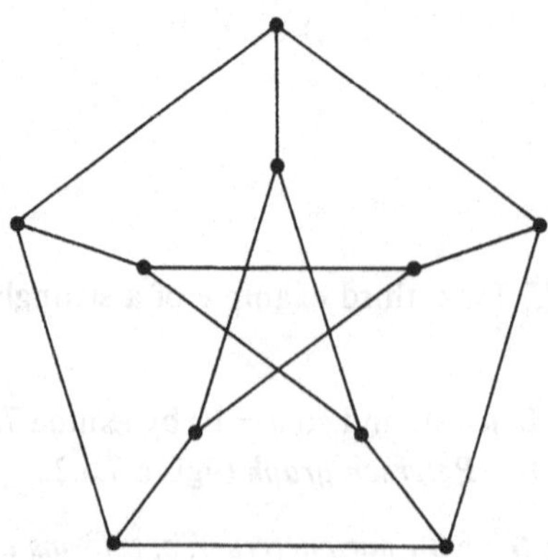

Figure 7.3.2.

strongly regular graph with parameters k, λ and μ satisfying the equations of lemma 7.3.2 for some integers α, s and t, $t \geq 1$, $s \geq 1$, $1 \leq \alpha \leq s+1$, $1 \leq \alpha \leq t+1$ (we call this a *pseudo-geometric graph*) it is not necessarily the case that the graph is the graph of a partial geometry. If such a graph *is* the graph of a partial geometry we shall call it *geometric*.

In deciding whether or not a pseudo-geometric graph is geometric, we try to reconstruct the lines of the partial geometry. In general the lines should correspond to maximal cliques (see section 5.2) of $s+1$ points. The best results on such constructions are due to Bose (1972). Cameron, Goethals and Seidel (1970) have shown that if $s = t^2$, or dually, then the graph is geometric (or see Seidel (1979)).

For the pentagon graph of example 7.3.1 above, we see that $\mu = \alpha(t+1) = 1$ implies $t = 0$ and $\alpha = 1$. Further, $k = s(t+1) = 2$ implies $s = 2$. The 'geometry' would look like figure 7.3.3 which satisfies GQ1, but not GQ2, and hence is not a partial geometry.

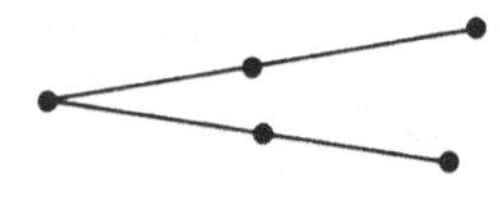

Figure 7.3.3.

For the Petersen graph we have the equations $3 = s(t+1)$, $0 = t(\alpha - 1) + s - 1$ and $1 = \alpha(t+1)$. The second equation implies $s = 1$. Thus $t = 2$ from the first equation. But now it is impossible to solve the last equation.

The *complement* of any graph $G = (V, E)$ is the graph G^c with point set V and edge set $E^c = \{(p, q) | (p, q) \notin E,\ p,\ q \in V\}$.

The complements of the pentagon and Petersen graphs are shown in figures 7.3.4(a) and (b) respectively. Note that the pentagon is its own complement.

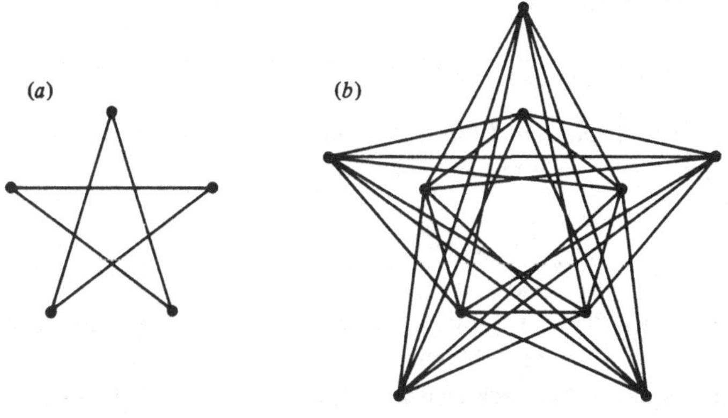

Figure 7.3.4.

Lemma 7.3.4. *The complement of a strongly regular graph with parameters k, λ, μ, and v points, is strongly regular with parameters $k' = v - k - 1$, $\lambda' = v - 2k + \mu - 2$ and $\mu' = v - 2k + \lambda$.*

The proof of this is left as a straightforward exercise for the reader.

Consider the partial geometry represented by a triangle which has $s = 1$. It is its own point graph. Its complement has no lines and therefore is not a partial geometry. Consider the partial geometry represented by figure 7.1.1. Its point graph is given in figure 7.3.5(*a*) and the complement of the point graph is given in figure 7.3.5(*b*). This complement clearly cannot be the graph of a partial geometry.

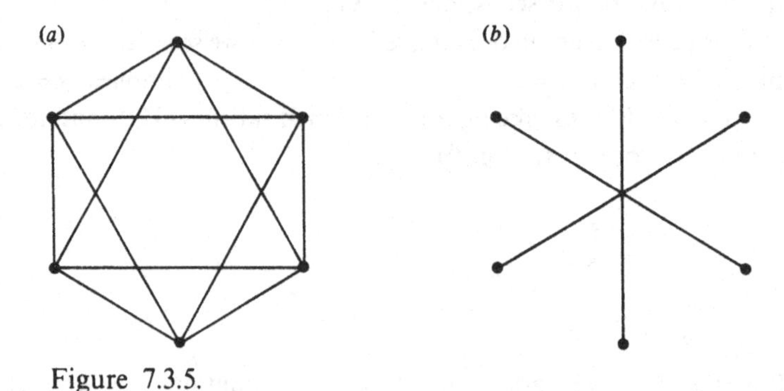

Figure 7.3.5.

We ask then, under what conditions the complement of the graph of a partial geometry is again the graph of a partial geometry. In fact this is a very difficult question. An easier one in light of the remarks about geometric graphs is to ask when the complement of the graph of a partial geometry is pseudo-geometric.

Lemma 7.3.5. *If there exists a partial geometry S with parameters α, s, t such that*

(1) $s > \alpha$,

(2) $\alpha | st$,

(3) $t(s - \alpha) \leq (s - \alpha + 1)\alpha$

then the complement G^c of the point graph G of S is a pseudo-geometric graph with parameters

$$\alpha' = \frac{t(s-\alpha)}{\alpha}, \quad s' = \frac{st}{\alpha}, \quad t' = s - \alpha.$$

Proof. It suffices to show that these parameters are integers satisfying the equations of lemma 7.3.2 with $t' \geq 1$, $s' \geq 1$, $1 \leq \alpha' \leq s' + 1$, $1 \leq \alpha' \leq t' + 1$. That is, we must check that $k' = s'(t' + 1)$, $\lambda' = t'(\alpha' - 1) + s' - 1$ and $\mu' =$

$\alpha'(t'+1)$ where k', λ' and μ' are the parameters of G^c. Thus, by lemma 7.3.4, $k' = v - k - 1$, $\lambda' = v - 2k + \mu - 2$ and $\mu' = v - 2k + \lambda$, where v is the number of points in S.

The fact that α' and s' are integers follows from condition (1). Clearly t' is always an integer. The fact that all are greater than 1 follows from condition (1) and the fact that α, s and t are themselves parameters of a partial geometry. Moreover, $\alpha' \leq s' + 1$ is trivial, and $\alpha' \leq t' + 1$ follows from (3).

The remaining computations to finish the proof are straightforward. □

Corollary. *If two of α', s', t' are the same as α, s, t, respectively, then they are all the same, and then $v = (2\alpha + 1)^2$, $s = 2\alpha$. $t = \alpha$.*

Proof. Once again, this is simply a matter of checking. □

Finally, we remind the reader that the above results also hold for the dual of a partial geometry.

7.4 Subgeometries

A (*partial*) *subgeometry* $S' = (P', L')$ of a partial geometry $S = (P, L)$ is a partial geometry such that $P' \subseteq P$ and L' is the restriction of L to P'. If $S' \neq S$, S' is said to be a *proper subgeometry.*

If $\alpha = 1$ in S then clearly $\alpha' = 1$ in S' for any subgeometry S' of S. Hence subgeometries of generalized quadrangles are just subquadrangles.

Example 7.4.1. Let $\Pi' = P(2, k')$ be a projective subplane of the projective plane $\Pi = P(2, k)$, $k' \leq k$. Let ℓ be a line of $P(2, k)$ which is also a line of $P(2, k')$. Let $W \subseteq \ell$ and $W' = W \cap S'$ where $|W'| \geq 2$. By example 7.1.4, the geometry with point set $\Pi' \backslash \ell$, and with line set defined using W, is a subgeometry of the geometry formed by taking the point set $\Pi \backslash \ell$ with lines defined using W.

Example 7.4.2. Let K be a $\{kd - k + d; d\}$-arc, $1 < d < k$ in the projective plane $\Pi = P(2, k)$ over GF(k). Embed Π as a plane in $P(3, k)$. Construct the partial geometry $S = (P, L)$ of lemma 7.2.2 with parameters $\alpha = d - 1$, $s = k - 1$ and $t = (k + 1)(d - 1)$. Let ℓ be a line of Π meeting the arc K in d points. Let $\Pi' = P'(2, k)$ be a second plane of $P(3, k)$ on ℓ. Then it is easy to see that the partial geometry S restricted to Π' is a partial geometry (a net in fact) with parameters $\alpha' = \alpha$, $s' = s$, $t' = d - 1$.

We now prove some results concerning subgeometries of partial geometries.

Lemma 7.4.1. *Let S be a partial geometry with parameters α, s, t, and let S' be a proper subgeometry with parameters $\alpha' = \alpha$, s', t'. Then $v \neq v'$ and $b \neq b'$.*

Proof. If $v = v'$, lemma 7.1.1 implies $(s+1)(st+\alpha)/\alpha = (s'+1)(s't'+\alpha)/\alpha$. Then $s \geq s'$ and $t \geq t'$ imply $s = s'$ and $t = t'$. Hence, $S = S'$, a contradiction. Similarly, $(t+1)(st+\alpha)/\alpha = (t'+1)(s't'+\alpha)/\alpha$ yields a contradiction. □

***Lemma* 7.4.2.** *Let S be a partial geometry with parameters α, s and t, and let S′ be a subgeometry with parameters $\alpha' = \alpha$, s', t'. If $p \in P \backslash P'$, then p is on at most one line of S′. If $p \in P \backslash P'$ is on a line of S′, then no other line on p meets S′.*

Proof. Let ℓ be a line of L' on $p \in P \backslash P'$. Let h be a second line on p with a point q in P'. Then q meets ℓ in S' in α' lines, while h is not one of those lines. Hence $\alpha \geq \alpha' + 1$, a contradiction. □

The result presented in the next theorem is a generalization of the result of theorem 6.5.1.

***Theorem* 7.4.3.** *If S is a partial geometry with parameters α, s, t, and S′ is a subgeometry of S with parameters $\alpha = \alpha', s, t$, then $s = s'$ or $s \geq s't' + \alpha - 1$, and dually $t = t'$ or $t \geq s't' + \alpha - 1$.*

Proof. Let $X = \{x \in P \backslash P' \mid x \text{ is not on any line of } L'\}$. We calculate $d = |X|$. The number of points in $P \backslash P'$ is

$$(s+1)\frac{(st+\alpha)}{\alpha} - (s'+1)\frac{(s't+\alpha)}{\alpha}.$$

The number of points of $P \backslash P'$ on lines of L' is $(t'+1)(s't'+\alpha)(s-s')/\alpha$. So

$$d = (s+1)\frac{(st+\alpha)}{\alpha} - (s'+1)\frac{(s't'+\alpha)}{\alpha} - (t'+1)\frac{(s't'+\alpha)(s-s')}{\alpha}$$
$$\geq 0.$$

If $t = t'$, then $(s+1)(st+\alpha) - (s'+1)(s't+\alpha) - (t+1)(s't+\alpha)(s-s') \geq 0$ which reduces to $(s-s')t(s-s't-\alpha+1) \geq 0$. Hence $s = s'$ or $s \geq s't' + \alpha - 1$.

Suppose $t > t'$. We count the ordered pairs (x_i, z) for $x_i \in X$, $z \in P'$ with $z \sim x_i$. There are $(s'+1)(s't'+\alpha)/\alpha$ choices for z and, for each of these, $(t-t')s$ choices for x_i. Letting t_i be the number of points of P' collinear with x_i, we have

$$\sum_{i=1}^{d} t_i = (s'+1)\left(\frac{s't+\alpha}{\alpha}\right)(t-t')s. \qquad (1)$$

Now count the ordered triples (x_i, z, z') for $x_i \in X$, z, $z' \in P'$ with $x_i \sim z$, $x_i \sim z'$ and $z \neq z'$. Again, there are $(s'+1)(s't'+\alpha)/\alpha$ choices for z. Then there are $(t'+1)s'$ choices of $z \sim z'$, and $v' - (t'+1)s' - 1 = s't'(s' + 1 - \alpha)/\alpha$ choices of $z \not\sim z'$. For $z \sim z'$, there are $(t-t')(\alpha-1)$ choices for x_i. For $z \not\sim z'$, there are $(t-t')\alpha$ choices for x_i. Thus

$$\sum_{i=1}^{d} t_i(t_i - 1) = \frac{(s'+1)(s't'+\alpha)}{\alpha}\left((t'+1)s'(t-t')(\alpha-1) + \frac{t's'(s'+1+\alpha)}{\alpha}(t-t')\alpha\right). \quad (2)$$

Adding (1) and (2) yields

$$\sum_{i=1}^{d} t_i^2 = \frac{(s'+1)(s't'+\alpha)}{\alpha}(t-t')(s'\alpha + t's'^2 + s - s').$$

By lemma 6.1.5, $d\sum_{i=1}^{d} t_i^2 \geq (\sum_{i=1}^{d} t_i)^2$ which, upon substituting the above, reduces to (check!)

$$(t-t')(s-s')[st + s't'(s't'+\alpha-1)][s-(s't'+\alpha-1)] \geq 0.$$

From $t \geq t'$ it follows that $s = s'$ or $s \geq s't' + \alpha - 1$. The remainder of the theorem follows by duality. □

Corollary. *If S is a partial geometry with parameters α, s, t, and S' is a subgeometry of S with parameters $\alpha = \alpha'$, s', t' then we have the following lower bound on v':*

$$\frac{s+1}{\alpha}(s'(t'+1) - (s-\alpha)) \leq v'.$$

Proof. By the last theorem, $[s-(s't'+\alpha-1)](s-s') \geq v$. Hence $(s'+1)(s't'+\alpha) - (s+1)[s't'+\alpha-s+s'] \geq 0$ and so $v' \geq [(s+1)/\alpha]$ $[s'(t'+1)-(s-\alpha)]$. □

The corollary is true even if $\alpha > \alpha'$. In addition it is possible to show that, even for this more general case, we have the following upper bound on v':

$$v' \leq \frac{st+\alpha}{\alpha(t+1)}(s'(t'+1)+t+1).$$

7.5 Pasch's axiom

In any near-linear space, *Pasch's axiom* states the following:

If ℓ_1 and ℓ_2 are any distinct lines intersecting in the point x, and if h_1 and h_2 are any distinct lines not on x and both meeting ℓ_1 and ℓ_2, then h_1 and h_2 intersect. (See figure 7.5.1.) (Exercise 53 of section 3.11 points out that if this axiom holds in a linear space, then the space is a projective space.)

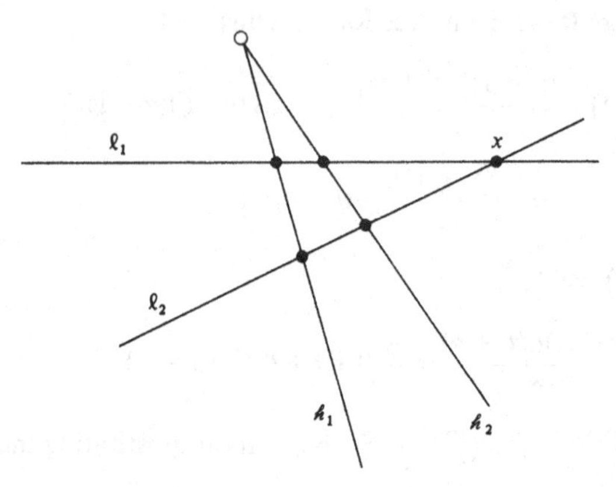

Figure 7.5.1.

Lemma 7.5.1. *If S is a partial geometry with parameters $\alpha = t + 1$, s and t, then Pasch's axiom holds.*

Proof. Let $p \in h_2 \backslash h_1$. Since $\alpha = t + 1$, every line on p meets h_1. In particular, h_2 does. □

If $\alpha = s + 1$, it is not necessarily the case that a partial geometry satisfies Pasch's axiom. In the example of figure 7.1.2, there are configurations which do not satisfy the requirement $h_1 \cap h_2 \neq \varnothing$. We leave it to the reader to find them.

In the case of a generalized quadrangle, Pasch's axiom is satisfied vacuously since, given intersecting lines ℓ_1 and ℓ_2, neither h_1 nor h_2 as described above can exist.

The dual of Pasch's axiom is called the *diagonal axiom* and is given by the following.

If p and q are any distinct points on a common line ℓ, and if x_1 and x_2 are points not on ℓ, each collinear with p and q, then x_1 and x_2 are collinear. (See figure 7.5.2.)

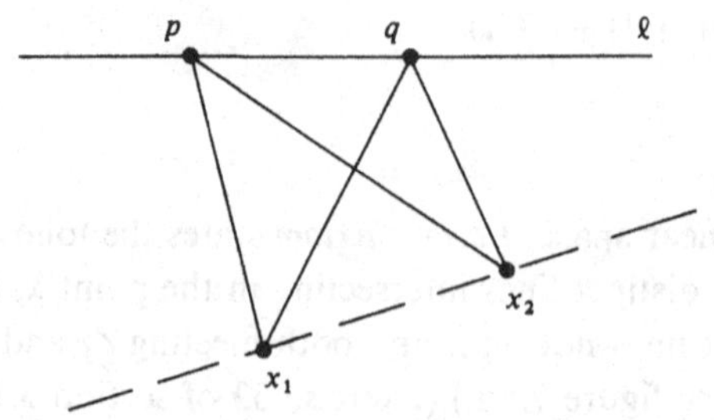

Figure 7.5.2.

Clearly, if a partial geometry satisfies Pasch's axiom, then the dual partial geometry satisfies the diagonal axiom and vice versa. We also have the next result which is merely the dual of lemma 7.5.1.

Lemma 7.5.2. *If S is a partial geometry with parameters $\alpha = s+1, s, t$, then the diagonal axiom holds.*

Lemma 7.5.3. *Let S be a partial geometry with parameters α, s and t satisfying the diagonal axiom and such that $a \neq 1$, $s+1$. Then S has proper subgeometries with parameters $\alpha' = \alpha$, $s' = \alpha - 1$, $t' = t$.*

Proof. Let p and q be distinct collinear points of S and define a substructure $S(p, q) = (P', L')$ by $P' = P_1 \cup P_2$ where P_1 is the set of $t(\alpha - 1) > 0$ (since $\alpha > 1$) points x for which $p \sim x \sim q$ and $x \notin \ell = pq$, and P_2 is the set of α points on ℓ collinear with some fixed point $r \in P_1$. As a consequence of the diagonal axiom, P_2 is independent of the choice of r. Let the set L' be the elements of L restricted to P'.

We show first of all that any two points of P' are adjacent. Suppose $x, y \in P_1$. Then x and y both adjacent to p and q implies by the diagonal axiom that $x \sim y$. Now let $x \in P_1$, $y \in P_2$. We may suppose $x \neq r$, $y \neq p$. Then x and y both adjacent to p and r and $p \sim r$ imply by the diagonal axiom that $x \sim y$.

Now let ℓ' be any line of L', $\ell' \neq \ell$. So there are points x and y of ℓ' in P'. (Note that $\ell' \cap \ell$ may be a point in P'. In fact this intersection could be x or y.) Let z be any point of $P' \setminus \ell$. By the above argument, $z \sim x, y$. Suppose $z \sim u$ where $u \neq x, y$ is any point of the 'extension' of ℓ' in S. Now $p, u \notin yz$ and p and u both adjacent to y and z imply by the diagonal axiom that $p \sim u$. Similarly $q \sim u$, and so $u \in P_1$. Hence all α points of ℓ' in L meeting z are in P'. (See figure 7.5.3.)

It follows also, since all points of P' are collinear, that each line of L' has α points.

Finally, to show that each point of P' is on $t+1$ lines in L', fix a point $x \in P_1$ and let $\ell' \ni x$. Since $\alpha > 1$ there is a point y of ℓ', $y \neq x$, adjacent

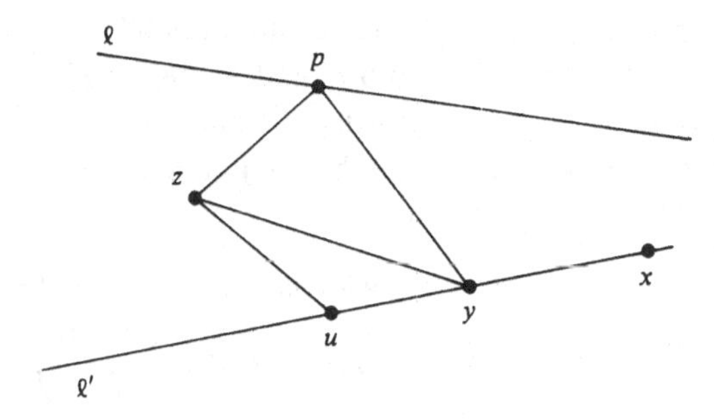

Figure 7.5.3.

to p. By the diagonal axiom, y is also adjacent to q and so is in P' or on ℓ. If $y \in \ell$, the fact that P_2 was independent of the choice of r also yields $y \in P'$.

If $x \in P_2$, $x \in \ell' \neq \ell$, then $\alpha > 1$ implies there is a point $y \in \ell'$, $y \neq x$ such that $r \sim y$. Without loss of generality, assume $x \neq q$, and use the diagonal axiom on x, q, r and y to yield $q \sim y$ and so $\ell' \in L'$. The fact that $\alpha \neq s+1$ implies that these subgeometries are proper. □

Corollary. *If S is as in lemma 7.5.3, then $(\alpha-1)|s, \alpha|t(t+1)$ and $s \geq (t+1)(\alpha-1)$. Moreover, if $\alpha \neq t+1$, then $\alpha|(s+1)$.*

Proof. The inequality is an immediate consequence of theorem 7.4.3, since $s \neq s'$.

To see that $(\alpha-1)|s$, consider the line ℓ of the lemma and fix the point p. The other s points of ℓ are partitioned into sets of size $\alpha-1$ according as we pick q and so determine $S(p,q)$. This gives the result. Finally, $\alpha|(t+1)ts'$ in *any* partial geometry with parameters α, s', t, by lemma 7.1.1. Here, this means $\alpha|t(t+1)(\alpha-1)$. Since α and $\alpha-1$ are relatively prime for $\alpha > 2$, we get $\alpha|(t(t+1)$ if $\alpha > 1$, as required.

Finally, suppose $\alpha \neq t+1$ in which case there exist non-intersecting lines ℓ_1 and ℓ_2 say. Let $x_i \in \ell_i$, $i = 1, 2$, $x_1 \sim x_2$. The set $S(x_1, x_2)$ contains α points of ℓ_1. Let $y_1 \in \ell_1 \backslash S(x_1, x_2)$. So $y_1 \not\sim x_2$. Let $y_2 \in \ell_2$ such that $y_1 \sim y_2$. The set $S(y_1, y_2)$ contains α points of ℓ_1. If there is a point z of ℓ_1 which is in both $S(x_1, x_2)$ and $S(y_1, y_2)$ then, by the diagonal axiom, $y_1 \sim x_2$, a contradiction. So we get a partition of the $s+1$ points of ℓ_1 into sets, each of size α. □

As we have already mentioned, most linear spaces satisfying Pasch's axiom can be shown to be projective spaces (exercise 53 of section 3.11). If we wanted to embed a partial geometry S with parameters α, s, t in a projective space of order s, then S must satisfy Pasch's axiom. Hence it is rather useful to have information about this property.

Let S be a partial geometry. For fixed intersecting lines ℓ and h, let $\alpha(\ell, h)$ be the number of (unordered) pairs of intersecting lines (ℓ', h'), ℓ' and h' both distinct from ℓ and h such that both ℓ' and h' meet both ℓ and h. Call a line $\bar{\ell}$ a *transversal* of ℓ and h if it is different from ℓ and h and meets ℓ and h in distinct points. Let $a(\ell, h)$ be the number of (unordered) pairs of intersecting transversals of ℓ and h.

Lemma 7.5.4. *Let S be a partial geometry with parameters α, s, t. Then, for fixed intersecting lines ℓ and h, $2s\binom{\alpha-1}{2} \leq a(\ell, h) \leq \binom{s(\alpha-1)}{2}$.*

Proof. Fix a point $p \in \ell \backslash h$. Then p is on $\alpha-1$ lines different from ℓ and meeting h. Any two of these form a pair of intersecting transversals. This is true also for each point of $h \setminus \ell$. Hence the first inequality. Note also that $a(\ell, h)$ is a maximum when any line meeting ℓ and h meets every other line meeting ℓ and h. The number of lines meeting ℓ and h not in $\ell \cap h$ is $s(\alpha - 1)$ and the number of ways in which two of these can be chosen is therefore $\binom{s(\alpha-1)}{2}$. □

Lemma 7.5.5. *Let S be a partial geometry with parameters α, s, t, Then, for fixed intersecting lines ℓ and h,*

$$\alpha(\ell, h) = \binom{t-1}{2} + a(\ell, h).$$

Proof. We note first of all that the intersection of distinct lines meeting ℓ and h may be $\ell \cap h$, a point of $\ell \setminus h$ or of $h \setminus \ell$, or a point of $P \setminus (\ell \cup h)$. In the first case, there are $\binom{t-1}{2}$ ways of choosing such pairs of lines. The other three cases make up the possibilities for $a(\ell, h)$. □

Corollary. *A necessary and sufficient condition for a partial geometry with parameters α, s, t to satisfy Pasch's axiom is that $\alpha(\ell, h) = \binom{t-1}{2} + \binom{s(\alpha-1)}{2}$ for all pairs of intersecting lines ℓ and h; equivalently $a(\ell, h)$ assumes the upper bound of lemma 7.5.4.*

Proof. Because of remarks in the proof of lemma 7.5.4, we see that a partial geometry satisfying this equation must indeed satisfy Pasch's axiom. Conversely, if Pasch's axiom holds, then the equation is satisfied. □

The following very deep theorem was proved by Sprague (1981). Unfortunately the background needed to understand the proof is beyond the scope of this text, but the enthusiastic reader might find this a very stimulating seminar topic.

A near-linear space is *connected* if its point graph is connected. (See section 5.5.)

Theorem 7.5.6. *Let $S = (P, L)$ be a finite connected near-linear space satisfying Pasch's axiom, such that $c(p, \ell) \neq 1, 2$ for any point p and line ℓ not on p, and such that the dual space of S satisfies Pasch's axiom. Then*

(i) *S is trivial (∅, a point, a line), or*
(ii) *S is a projective plane, or*
(iii) *S is isomorphic to the set of i-dimensional and $(i+1)$-dimensional*

subspaces of a d-dimensional vector space, $d \geq i$, $1 \leq 1 \leq d$, over a finite field $GF(k)$, k a prime power.

7.6 A history of partial geometries

As has been mentioned, Bose was the first to introduce the concept of a partial geometry in 1963. He pointed out that there were already many connections between these new systems and the structures of known systems such as designs (Dembowski, 1968), nets (Bose, 1963), association schemes (Bose, 1963) and of course generalized quadrangles and strongly regular graphs.

The study of partial geometries has generally been divided into two parts. On one hand there is the search for combinatorial properties; on the other hand there is the search for connections with projective spaces. (This follows the usual pattern of investigations in geometry: combinatorial versus structural.)

Most of the combinatorial work has been done by, apart from Bose, Thas (1977*a*, *b*; 1978*b*) and De Clerck (1978, 1979*b*). One of the most difficult combinatorial questions is a problem we mentioned in section 7.3: when is a pseudo-geometric graph geometric? In his original paper, Bose gave a sufficient condition for this:

Theorem. *A pseudo-geometric graph with parameters α, s, t is geometric if*

$$s + 1 > \tfrac{1}{2}[(t+1)t + \alpha(t+2)(t^2+1)].$$

De Clerck (1979*b*) has a classification of the pseudo-geometric graphs in case $\alpha = s - 1$:

Theorem. *If G is a pseudo-geometric graph with parameters α, s, t, then one of the following occurs.*

(1) $s = 2, t = 1, 2$ *or* 4, *G is geometric and the corresponding generalized quadrangle is unique.*
(2) $s = 3$, $t = 1$, 2, *or* 4. *If* $t \neq 4$ *then G is never geometric.*
(3) $s > 3$, $t = s - 1$ *and G is geometric if and only if there exists a projective plane of order* $s + 1$.
(4) $s > 3, t = 2(s-1)$ *and G is geometric if there exists a complete oval in some projective plane of order* $2s$.

The work on the structure of partial geometries is due mainly to Thas and De Clerck (De Clerck and Thas, 1978; Thas and De Clerck, 1975, 1977). The standard approach has been to try to embed a given partial geometry in a projective space. Thas (1978*a*) introduced a variation on this approach by trying to embed the partial geometry in an affine space.

To date, partial geometries have been generalized in two directions. The concept of *partial geometry of dimension 3* was introduced by Laskar and Dunbar (1978), and further work has been done on it by Thas (1980*b*). Essentially, partial geometries of dimension 3 have the same type of structure as partial geometries, but they properly contain as 'hyperplanes' partial geometries. (See exercise 42 of section 7.7.)

The other generalization of partial geometry is to *semi-partial geometry* where we allow $c(p, \ell)$ to be either 0 or a fixed α for each point p not on a line ℓ. In addition, if two points are not collinear, we suppose that there is a fixed positive number of points collinear with both. For results on these structures, see the relevant papers in the bibliography by Debroey and Thas.

7.7 Exercises

1. Show that if S is a non-empty partial geometry with $\alpha = 1$, then S is a generalized quadrangle.
2. Construct a partial geometry with parameters $\alpha = s = 2$, $t = 3$.
3. Show that the Shrikhande geometry of example 7.1.3 cannot be embedded in an affine plane of order 4.
4. Show that if $\alpha = t$ in a partial geometry $S = (P, L)$, then S satisfies the 'parallel postulate': given any point p not on a line ℓ, there is a unique line on p missing ℓ.
5. Under the conditions of exercise 4, show that parallelism is an equivalence relation on the lines of L.
6. $A(3, 2)$, affine 3-space of order 2, falls into subfamily (*a*) of the family of partial geometries with $\alpha = s + 1 = 2$. Construct the dual of this partial geometry.
7. Prove lemma 7.1.2.
8. Check the details in example 7.1.5 to show that the construction really yields a partial geometry.
9. Construct H_2^3 giving the points and lines, and drawing a diagram if possible.
10. Show that if S is a proper partial geometry with parameters α, $s = t = 5$, then $\alpha = 2$.
11. If S is a partial geometry with $s = t$ and $\alpha = 2$, show that $s = 1$, 2 or 5.
12. Find a maximal arc in the projective plane $P(2, 4)$.
13. Show that the converse of lemma 7.3.2 is not necessarily true. That is, a pseudo-geometric graph is not always geometric.
14. Decide whether or not the Petersen graph represents a partial geometry.

15. For a grid with parameters $s = 2$, $t = 1$ draw the corresponding strongly regular point graph. Do the same for the dual grid.
16. Draw the corresponding strongly regular point graph for the generalized quadrangle of example 6.1.3.
17. For the partial geometry of example 7.1.2 draw the point and line graphs.
18. Prove that if a partial geometry has $s = 1$ then it is isomorphic to its point graph.
19. If the point and line graphs of a partial geometry S are isomorphic, what can you say about S?
20. Prove lemma 7.3.4.
21. Let V be a set of $n \geq 5$ elements. The *triangular graph* $T(n)$ is the graph whose vertices are the unordered pairs from V, two vertices being adjacent whenever the corresponding pairs have an element in common. Prove that $T(n)$ is strongly regular with parameters $k = 2(n-2)$, $\lambda = n-2$, $\mu = 4$.
22. In exercise 21, what happens if $n < 5$?
23. Show that the Petersen graph is the complement of $T(5)$.
24. A Steiner system $S(t, k, v)$ is called a *Steiner triple system* if $t = 2$ and $k = 3$. Show that the set of blocks of a Steiner triple system with $v > 9$ forms a strongly regular graph with adjacency defined by $B_1 \sim B_2$ if and only if $B_1 \cap B_2 \neq \emptyset$.
25. Let S be a partial geometry with parameters α, s and t. Let P be the adjacency matrix of the point graph (i.e., the (i, j)th entry is 1 if the point p_i is adjacent to the point p_j, and 0 otherwise), and L the adjacency matrix of the line graph. Let N be the (point-line) incidence matrix. Show that $NN^t = P + (t+1)I$ and $N^tN = L + (s+1)I$, where I is the unit matrix.
26. Choose a subplane $\Pi' = P(2, 2)$ of $\Pi = P(2, 2^2)$, and a line ℓ of Π which is also a line of Π'. Choose a subset W of ℓ, with $|W| = 4$. Check that, as in example 7.4.1, this gives rise to a proper subgeometry of a partial geometry.
27. If S' is a subgeometry of S and both have parameters α, s, t show that $S = S'$.
28. If $S = (P, L)$ and $S' = (P', L')$ are partial geometries with the same parameters and such that $P' \subseteq P$, show that it is not necessarily the case that $L' = L$.
29. Show that if S' is a proper subgeometry of S where $s = s'$ and $t = t'$, then $v \neq v'$ and $b \neq b'$.
30. In the proof of theorem 7.4.3, show that $d\sum_{i=1}^{d} t_i^2 = (\sum_{i=1}^{d} t_i)^2$ implies $t_i = s't' + \alpha$ for all $1 \leq i \leq d$.
31. Check that $(t-t')(s-s')[st + s't'(s't' + \alpha - 1)][s - (s't' + \alpha - 1)] \geq 0$ in the proof of theorem 7.4.3 if $t > t'$.

32. Prove that the partial geometry of figure 7.1.2 does not satisfy Pasch's axiom.

33. Do the partial geometries of example 7.1.4 satisfy Pasch's axiom?

34. Show that H_k^d satisfies Pasch's axiom.

35. Prove lemma 7.5.2 directly.

36. Test the partial geometries of exercises 32, 33 and 34 for the diagonal axiom.

37. Compute $a(\ell, \hbar)$ and $\alpha(\ell, \hbar)$ for fixed intersecting lines ℓ and $\hbar$ of H_k^d.

38. Show that the values $a(\ell, \hbar)$ and $\alpha(\ell, \hbar)$ in a partial geometry are dependent on the intersecting lines ℓ and $\hbar$ chosen.

39. Show that it is possible that a near-linear space satisfies Pasch's axiom. while its dual space does not satisfy Pasch's axiom.

40. If S is a partial geometry with parameters $\alpha = s - 1, s, t$, show that if $s = 2$ or 3 then $t = 1$, 2 or 4.

41. If S has parameters α, s, t, show that $s-1 \mid 2t$.

42. Determine whether or not partial geometries have dimension 2. (See section 7.6.)

****43.** Find a definition of partial geometry which is similar to the definition given for generalized quadrangle. Your new definition may include some 'trivial' structures not included in the definition we give.

44. A *net* of degree r (Bose, 1963) is a system $S = (P, L)$ of points and lines such that P is non-empty, and the elements of L can be partitioned into r disjoint non-empty 'parallel classes' in such a way that each point of the net is incident with exactly one line of each class, and given two lines belonging to distinct classes, there is exactly one point of P incident with both lines. Show that a net is a partial geometry with parameters $\alpha = r-1$, $s = k-1$, $t = r-1$, for some k.

8

Blocking sets

Lyke as a huntsman after weary chace,
Seeing the game from him escapt away,
Sits downe to rest him in some shady place,
With panting hounds beguiled of their prey:
So, after long pursuit and vaine assay,
When I all weary had the chace forsooke,
The gentle deare returned the selfe-same way, . . .

Edmund Spenser *Amoretti* Sonnet LXVII

The concept of 'blocking' in a mathematical sense, seems to have been around for decades. Work in the early 1900's was in the context of topology and set theory and so dealt with infinite sets. See, for instance Bernstein (1908) and Miller (1937). In the 1950's and 1960's, a number of people independently introduced the idea for finite systems. Some of these people were interested in the application to game theory. Other applications have since been introduced in statistics and coding theory. We talk about these applications in section 8.6.

8.1 Definition and examples

Let $S = (P, L)$ be a near-linear space. A *blocking set* in S is a subset B of P such that for each line $\ell \in L$, $0 < |\ell \cap B| < v(\ell)$.

Example 8.1.1. Consider figure 1.2.1 of chapter 1. Blocking sets are $\{3, 4\}$, $\{1, 2, 5\}$. In fact it is easy to see that these are the only blocking sets. Note that if we restrict the space to $P' = \{1, 2, 3, 4, 5\}$, the set of all points on at least one line, then the blocking sets above are the complements of each other in P'.

Example 8.1.2. Figure 1.4.1 has $\{1, 3, 4, 6\}$ as a blocking set. However, this is no longer a blocking set in the linear space of figure 2.1.1. (See exercise 8.7.4.)

Example 8.1.3. The Fano plane has no blocking set; nor do $A(2, 2)$, $A(2, 3)$. All other affine and projective planes contain blocking sets. (See exercise 8.7.5.) It is not difficult to show that $P(2, 3)$ has precisely two blocking sets; this can be done by looking at cases and noting that some line must have precisely three points in the blocking set.

Example 8.1.4. The generalized quadrangle of figure 6.1.3 has $\{1, 2, 5, 6, 8, 10, 11, 14\}$ as one of its blocking sets.

In much of the work in this book, spaces have been equipped with more than two dimensions. Several definitions of 'blocking set' have arisen in the literature to take this into account. We mention some of the definitions below but give no results.

Bruen (1980) and Bruen and Thas (1982) define a blocking set of $P(d, q)$ to be a set of points X such that each hyperplane meets X, but X contains no line.

Elbert (1978) bases his definition on the concept of a 'winning coalition' found in Richardson (1956): a blocking set in $P(d, q)$, d even, is a set of points X which meets each but does not contain any $d/2$-subspace; similarly for d odd and $(d+1)/2$-subspaces.

In Mazzocca and Tallini (1985), Beutelspacher and Eugeni (1986) and in Beutelspacher and Mazzocca (1987), a *t-blocking* set in $P(d, q)$ or in $A(d, q)$ is defined to be a subset X of points of the space such that every t-space meets X and its complement.

Finally, we mention that in Beutelspacher (1980), a *t-blocking* set in $P(d, q)$, $d \geq t+1$, is defined to be a set of points X such that any $(d-t)$-space meets X and no $(d-t)$-space is contained in X.

The reader can check that each of these definitions is, in one way or another, a generalization of our definition. Unless otherwise indicated, the term 'blocking set' throughout the remainder of this chapter will refer to the definition given in the first paragraph of this section.

The basic questions considered in this chapter are, for a given near-linear space or family of near-linear spaces: do blocking sets exist? how many points do they have? what do they look like? We will not always be able to answer all of these questions.

8.2 Blocking sets in projective planes

The 'classical' results on blocking sets are in the context of projective planes and are due to Bruen (1970, 1971). A new proof was published by Bruen and Silverman in 1987 and this is the one we give here. In the proof, we use the following facts which are left as an exercise. If a projective plane π has finite order n, and if B is a blocking set in π, then $|B| = n + \lambda$ for some $\lambda \geq 1$. Moreover, no line contains more than λ points of B.

Theorem 8.2.1. *Let B be a blocking set in a finite projective plane π of order n. Then*

$$n + \sqrt{n} + 1 \leq |B| \leq n^2 - \sqrt{n}.$$

Equality holds if and only if B is the set of points of a proper Baer subplane or its complement.

Proof. Let x_i, $1 \le i \le k$, denote the number of lines of π containing exactly i points of B, where k is the maximum size of line intersections with B. Thus

$$\sum_{i=1}^{k} x_i = n^2+n+1. \tag{A}$$

Counting incidences of lines of π first with points of B, then with ordered pairs of points of B, we obtain, letting $|B| = n+\lambda$,

$$\sum_{i=1}^{k} x_i i = (n+\lambda)(n+1), \tag{B}$$

$$\sum_{i=1}^{k} x_i i(i-1) = (n+\lambda)(n+\lambda-1). \tag{C}$$

Forming $-\lambda(\mathrm{A})+\lambda(\mathrm{B})-(\mathrm{C})$ produces

$$\sum_{i=1}^{k} x_i(i-1)(\lambda-i) = \lambda^2 n - 2\lambda n - (n^2-n).$$

Since the left-hand side is non-negative (no line of π contains more than λ points of B), so is the right-hand side.

Since $n > 0$, we have

$$\lambda^2 - 2\lambda - (n-1) \ge 0,$$

that is,

$$[\lambda-(\sqrt{n}+1)][\lambda+(\sqrt{n}-1)] \ge 0.$$

Since $\lambda > 0$, we obtain $\lambda \ge \sqrt{n}+1$ with equality if and only if $x_i = 0$ for all i save $i=1$ and $i=\lambda=\sqrt{n}+1$. Using the fact that the complement of B is also a blocking set, the rest of the theorem follows easily. □

Can blocking sets in projective planes take on sizes corresponding to *all* values in the range specified by theorem 8.2.1? The following results show that in finite Desarguesian planes the answer is 'almost' yes. They are due to Berardi and Eugeni (1984a).

Lemma 8.2.2. *Let $n = p^h \ge 3$, p a prime. Then for any integer s satisfying $2n \le s \le n^2-n+1$ there exists a blocking set in $P(2,n)$ with precisely s points.*

Proof. We note first that for the case $n=3$, the above statement is not difficult to verify. See example 8.1.3.

Suppose that n is an odd prime power, $n > 3$. The plane π contains a non-degenerate conic C (a non-degenerate quadric!) by section 5.3. Let q be a point of C and ℓ the tangent to C at q. (See exercises 5.8.26 and 5.8.27.) It is easy to check that $(C \cup \ell) \setminus \{q\}$ is a blocking set on $2n$ points. Now the number of lines meeting $C \setminus \{q\}$ in two points is $\binom{n}{2} = n(n-1)/2$. Therefore each point of $\ell \setminus \{q\}$ is on an average of $n(n-1)/2n = (n-1)/2$ of these lines. It follows that some point of $\ell \setminus \{q\}$ is on at least $(n-1)/2$ lines not meeting C at all. On each of $(n-1)/2$ of these lines, take between 0 and $n-1$ points and adjoin them to $(C \cup \ell) \setminus \{q\}$ to form a set B with between $2n$ and $(n+1)^2/2$ points; B is again a blocking set (using $n > 3$). Counting possible orders of the complement of B now gives us blocking sets of all orders between $2n$ and $n^2 - n + 1$. $\square$

In fact, Berardi and Eugeni were, except for $n = 3$ for which it is false, able to include the values $2n-1$ and $n^2 - n + 2$ in this range. The proof above is a modification of theirs.

Theorem 8.2.3. *Let $n > 2$ be an integer. Then for any integer s satisfying $n + m(n) + 1 \le s \le n^2 - m(n)$, there exists a blocking set in $P(2, n)$ with precisely s points, where $m(n)$ is defined as follows:*

$$m(n) = \begin{cases} \sqrt{n} & \text{if } n \text{ is a square,} \\ (n+1)/2 & \text{if } n \text{ is a prime,} \\ p^{h-d} & \text{if } n = p^h,\ p \text{ a prime, } h \text{ odd and} \ge 3, \text{ and } d \\ & \text{is the greatest proper integer divisor of } h. \end{cases}$$

Proof. We begin with the case n square. Let π_0 be a (proper) Baer subplane of $\pi = P(2, n)$ and let $p \in \pi_0$. (Note that π_0 exists since π is Desarguesian.) The number of lines of π meeting π_0 precisely in p is $n - \sqrt{n}$. For any integer s, $0 \le s \le n - \sqrt{n} - 1$, choose one point, other than p, on precisely s of these lines. Let X be the set of these s points and define $B = \pi_0 \cup X$. Then B is a blocking set of π of size $n + \sqrt{n} + 1 + s$ where $n + \sqrt{n} + 1 \le n + \sqrt{n} + 1 + s \le 2n$. Using the complement of B in π yields blocking sets covering all values between and including $n^2 - n + 1$ and $n^2 - \sqrt{n}$. The remaining values can be obtained from lemma 8.2.2 above.

Now suppose that n is a prime. If $n = 2$, there are no values in the interval; so we may suppose $n \ge 3$. We construct once again a blocking set having the required number of points. This construction is due to Bruen (1971).

Consider the set $S = \{x^2 \mid x \text{ is in } \mathrm{GF}(n)\}$ and let

$$X = \{[0, s, 1], [-s, 0, 1], [1, s, 0] \mid s \in S\} \cup \{[0, 1, 0]\}$$

be a subset of the point set of $P(2, n)$. Then $|X| = n + (n+1)/2 + 1$, and we leave as exercise 8.7.12 that X is a blocking set in π.

The number of lines through the origin with slope an element of $S\backslash\{0\}$ is $(n-1)/2$. (See section 4.6.) Fix on each of these lines ℓ an arbitrary number x_ℓ of points of $\pi\backslash\ell_\infty$, with $0\le x_\ell\le n-2$. (Note that ℓ contains precisely two points of X in π.) Adjoining these points to the set X, we obtain a set B of points satisfying $n+(n+1)/2+1\le|B|\le(n^2+5)/2$, which is again a blocking set. Using $\pi\backslash B$, we also cover the values between and including $(n^2+2n-3)/2$ and $n^2-(n+1)/2$. Lemma 8.2.2 gives us the remaining values.

Finally, consider the case $n=p^h$, p a prime, h odd, $h\ge 3$, and let d be the greatest proper integer divisor of $h=md$. Letting $\Gamma=\mathrm{GF}(n)$ and $\Delta=\mathrm{GF}(p^d)$, we can consider Γ as an m-dimensional vector space over Δ. In fact, letting u be a fixed element of $\Gamma\backslash\Delta$, then $\{1,u,\dots,u^{m-1}\}$ is a basis for Γ over Δ. Define

$$I=\{a_1u+\cdots+a_{m-1}u^{m-1}\mid a_i\in\Delta\},$$

and let

$$B_1=\{[a,i,1]\mid a\in\Delta,\ i\in I\}\cup\{[1,i,0]\mid i\in I\}\cup\{[0,1,0]\}.$$

The set B_1 has order $p^h+p^{h-d}+1$. We claim that as a set of points of π, B_1 forms a blocking set. To see this, we use the co-ordinatization of section 4.6. The point $[0,1,0]$ is on ℓ_∞ and on all lines of the form $x=b$. If $y=mx+b$, $m\in I$ implies $[1,m,0]\in\langle -m,1,-b\rangle$; $b\in I$ implies $[0,b,1]\in\langle -m,1,-b\rangle$. The reader is left to complete the detail that no line is entirely contained in B_1.

Now each of the $p^d-1\ge 1$ lines with equation $y=ax$, $a\in\Delta\backslash\{0\}$ is tangent to the set B_1 at $[0,0,1]$. Fix one of these lines, say ℓ.

Suppose $p^d-1>1$. Then on each of the above tangent lines, *except* for ℓ, choose any number of points between 0 and p^h-1. Adjoin all of these points to B_1, resulting in a set B with

$$p^h+p^{h-d}+1\le|B|\le p^h+p^{h-d}+1+(p^h-1)(p^d-2).$$

We claim that B is again a blocking set. For lines of π on $[0,0,1]$, it is fairly easy to see that they both miss and meet B. Any line not on $[0,0,1]$ meets B_1 but also misses ℓ. Considering B and its complement, we cover the range of values $p^h+p^{h-d}+1$ to $p^h+p^{h-d}+1+(p^h-1)(p^d-2)$ and $p^{2h}-p^{h-d}-(p^h-1)(p^d-2)$ to $p^{2h}-p^{h-d}$. Some elementary computation (left as an exercise!), along with lemma 8.2.2, establishes the result we want.

We now only need to consider the case $p^d-1=1$; that is, $p=2$ and $d=1$. Here, ℓ is the unique tangent to B_1 having non-zero slope in Δ. Take between 0 and 2^h-1 points of ℓ and adjoin these to B_1 to obtain the set B.

Each line on $[0,0,1]$ meets B and its complement. Now there are other lines tangent to B_1 at $[0,0,1]$; for instance, $y=(i+1)x$ for any $i \in I$. All other lines of π meet B *and* meet a line of the above form, and so B is a blocking set in this case too. □

There are several results improving Bruen's bounds in special cases. We shall just mention a paper by Blokhuis and Brouwer (1986) which, for n odd, non-square and at least 7, improves the lower bound to $n+\sqrt{2n}+1$ in $P(2,n)$. For n square, Bruen's bound is, of course, the best one can do in $P(2,n)$. For n a prime, Blokhuis (1994) proves that the Berardi–Eugeni bound of $3(n+1)/2$ is the best one can do.

The statements of the above theorems consider the existence and size of blocking sets, but the proofs give explicit constructions so that we can see what some of the blocking sets look like. However, it is very likely that we have not constructed *all possible* blocking sets. Much research is still going on on this question.

8.3 Blocking sets in affine planes

We parallel the results of the previous section by first of all establishing bounds on blocking sets in arbitrary affine planes.

As pointed out in section 8.1, the affine planes of orders 2 and 3 have no blocking set. The first result of this section gives bounds for all $n \geq 4$; it is due to Bruen and Silverman (1987). As in the projective plane case, we argue that if $|B|=n+\lambda$ for a blocking set B in the affine plane A, then no line of A contains more than λ points of B. (Exercise 8.7.17.)

Theorem 8.3.1. *Let B be a blocking set in a finite affine plane A of order $n \geq 4$. Then $|B| > n+\sqrt{n}+1$ and, if n is a square, $|B| \geq n+\sqrt{n}+2$.*

Proof. Exercise 8.7.16 yields a blocking set in A on $2n-1$ points if $n \geq 5$. For $n=4$, $2n$ is the best possible. Therefore, we may set $|B|=n+\lambda$ with $\lambda \leq n$. Let I denote the set of incidences of ordered pairs of points of B with lines of A. Then $|I|=(n+\lambda)(n+\lambda-1)$.

Let C be any parallel class of A. We count the contribution I_C to I coming from the lines of C. Each of these n lines has at least one point in B, thus accounting for at least n points of B, and leaving us to distribute up to λ points on these lines in such a way that no line gets more than λ points in total. In doing this, when do we maximize the number of pairs of points on lines of C? Note that for x and y positive integers, $\binom{x+y}{2}$ is significantly bigger than $\binom{x}{2}+\binom{y}{2}$. (See exercise 8.7.18.) Hence, clustering the maximum number of points possible on one line of C maximizes the number of point pairs. Since no line can have more than λ points, we get

$$|I_C| \leq \lambda(\lambda-1)+2.$$

Thus,

$$|I| \leq (n+1)[\lambda(\lambda-1)+2].$$

Combining this with the value of $|I|$ above yields

$$0 \leq n\lambda^2 - 3n\lambda + 3n - n^2 + 2.$$

Dividing by n gives

$$-2/n \leq \lambda^2 - 3\lambda + (3-n)$$

which, since $n \geq 4$, implies

$$0 \leq \lambda^2 - 3\lambda + (3-n).$$

Treating this last inequality as a quadratic in λ yields $\lambda \geq (3+\sqrt{4n-3})/2$. Since $|B| = n+\lambda$ and $(3+\sqrt{4n-3})/2 > \sqrt{n}+1$, we get $|B| > n+\sqrt{n}+1$. In case n is a square, we can then say $|B| \geq n+\sqrt{n}+2$. □

In Brouwer and Schrijver (1978) (and also in Jamison (1977)) it is proved that a set of points of $A(2,n)$ which has a non-empty intersection with each line of the plane must have order at least $2n-1$. Note that if we add the condition that the set contain no line of $A(2,n)$, i.e. that it be a blocking set, the lower bound $2n-1$ is still operable. In view of exercise 8.7.16, it is a 'best' lower bound in case $n \geq 5$. Hence, we state the next theorem without proof, but refer the reader to Brouwer and Schrijver above. (The proof is short but not at all geometrical!)

Theorem 8.3.2. *Let B be a blocking set in $A(2,n)$, $n \geq 2$. Then $2n-1 \leq |B|$.*

Thus in the plane $A(2,n)$, $n \geq 5$, the possible range of values taken on by $|B|$ is $2n-1$ to $(n-1)^2$. We give next a result of Berardi and Eugeni (1984b) showing that each of these values can in fact occur in *any* affine plane of order at least 5.

Theorem 8.3.3. *Let A be an affine plane of finite order $n \geq 5$. Then for each integer s with $2n-1 \leq s \leq (n-1)^2$, there exists a blocking set in A with order s.*

Proof. Fix the points p_0, p_1, p_2, p_3 and the lines $\ell_1 = p_0p_1$, $\ell_2 = p_0p_2$, $\ell_3 = p_0p_3$, $\ell_4 = p_2p_3$, $\ell_5 = p_1p_3$, $\ell_6 = p_1p_2$ such that $\ell_1 \parallel \ell_4$ and $\ell_2 \parallel \ell_5$. Fix a point $p \in \ell_6 \setminus \{\ell_3 \cap \ell_6, p_1, p_2\}$. (Note that $\ell_3 \cap \ell_6$ may be the empty set.) Then it is easy to check that $B_0 = (\ell_1 \cup \ell_2 \cup \{p, p_3\}) \setminus \{p_1, p_2\}$ is a blocking set in A on $2n-1$ points.

Now fix a line ℓ_7 parallel to ℓ_1, but different from ℓ_1 and ℓ_4, and not on p or on $q=\ell_3\cap\ell_6$ if q exists. Consider a point $r\in p_3p_4\backslash\{p_3,p_4\}$, where $p_4=\ell_2\cap\ell_7$, and such that r, p_2 and $\ell_5\cap\ell_7$ are not collinear. Define

$$X_0=A\backslash(\ell_1\cup\ell_2\cup\ell_5\cup\ell_7\cup\{p,r\}).$$

Thus, the set complement of $B_0\cup X_0$ is

$$(B_0\cup X_0)^c=B_0^c\cap X_0^c=\ell_5\backslash\{p_3\}\cup\ell_7\backslash\{p_4\}\cup\{p_2,r\}$$

with order $2n-1$, and so $|B_0\cup X_0|=(n-1)^2$. In addition, $B_0\cup X_0$ is again a blocking set in A; we leave the details for the reader. Finally, to complete our claim, let X be a subset of X_0. It is easy to see that $B_0\cup X$ is still a blocking set and that it takes on all values s in the range $[2n-1,\,(n-1)^2]$ as the size of X varies. □

We also leave the reader the task of determining why the set B_0 above is not a blocking set in case $n=4$. (Exercise 8.7.21.)

The proof of theorem 8.3.3. gives us constructions of blocking sets, as does exercise 8.7.22 in the square order case.

8.4 Blocking sets in Steiner systems

Affine and projective planes are special examples of Steiner systems. Although these systems were defined earlier (section 2.1), we reexamine the definition here.

A Steiner system $S=S(2,k,v)$ is a finite linear space on v points such that each line is on a constant number k of points. By lemma 2.3.3, S is point regular with regularity $r=(v-1)/(k-1)$.

Example 8.4.1. A complete graph on $v\geq 2$ points is a Steiner system with line size $k=2$. It is easy to see that a blocking set can only exist when $v=2$.

Example 8.4.2. In figure 8.4.1 (overleaf) are two examples of Steiner systems on 25 points with line size 4. (Note that neither is an affine or projective plane!) The first has blocking set $\{1,2,3,4,5,6,13,14,15,16,17,18\}$. The second has no blocking set.

Constructing blocking sets in general Steiner systems can be very difficult as we have no systematic way of handling their structure as we did for affine and projective planes. We can, however, still produce some results pertaining to bounds on blocking set sizes. The following theorem is due to Drake (1985). Note that for $k=2$, $v>2$ or for $k=3$, $v>4$, the value under the root sign is negative.

1 2 3 25	2 6 10 19	3 9 13 14	5 14 18 19	9 12 15 25
1 4 16 24	2 7 20 21	3 10 12 22	6 7 8 24	10 14 16 21
1 5 12 21	2 8 13 15	4 5 6 25	6 9 11 21	11 15 17 19
1 6 13 22	2 3 16 18	4 7 12 19	6 12 14 17	12 13 18 20
1 7 14 15	2 11 12 24	4 8 9 22	6 15 16 20	13 19 23 24
1 8 17 18	3 4 11 20	4 10 15 18	7 10 13 25	14 20 22 24
1 9 19 20	3 5 15 24	4 13 17 21	7 11 18 22	15 21 22 23
1 10 11 23	3 6 18 23	5 7 9 23	8 11 14 25	16 19 22 25
2 4 14 23	3 7 16 17	5 8 10 20	8 12 16 23	17 20 23 25
2 5 17 22	3 8 19 21	5 11 13 16	9 10 17 24	18 21 24 25

1 2 4 25	1 3 8 11	1 5 9 24	1 6 7 15	1 10 13 17
1 12 19 21	1 14 18 22	1 16 20 23	2 3 14 19	2 5 8 16
2 6 9 12	2 7 18 23	2 10 11 22	2 13 15 20	2 17 21 24
3 4 10 15	3 5 7 13	3 6 21 25	3 9 16 22	3 12 17 23
3 18 20 24	4 5 11 20	4 6 8 14	4 7 16 21	4 9 17 18
4 12 13 24	4 19 22 23	5 6 17 22	5 10 21 23	5 12 14 15
5 18 19 25	6 10 19 20	6 11 23 24	6 13 16 18	7 8 19 24
7 9 10 25	7 11 14 17	7 12 20 22	8 9 15 23	8 10 12 18
8 13 21 22	8 17 20 25	9 11 13 19	9 14 20 21	10 14 16 24
11 12 16 25	11 15 18 21	13 14 23 25	15 16 17 19	15 22 24 25

Figure 8.4.1.

Theorem 8.4.1. *Let B be a blocking set in a Steiner system with v points, b lines, point size r and line size k. Then*

$$\frac{v}{2}-\frac{1}{2k}\sqrt{v^2k^2-4v^2k+4vk}\le |B| \quad \text{if } k>3.$$

Proof. Let $w=|B|$. Let x and y be integers satisfying $w-1=(k-2)x+y$ where $0\le y\le k-3$. (Why do they exist?) Now

$$b=\sum_{\ell\in L}1=\sum_{\ell\in L}\sum_{p\in(B\cap\ell)}\frac{1}{|B\cap\ell|}=\sum_{p\in B}\sum_{\ell\,\text{on}\,p}\frac{1}{|B\cap\ell|}.$$

We estimate $\sum_{\ell\,\text{on}\,p}(1/|B\cap\ell|)$. Using the above equation in x and y, we see that $x+y\le r$ (counting w by fixing $p\in B$). In order to maximize the above sum, we put as many 1's as possible in it. Therefore

$$\sum_{\ell\,\text{on}\,p}\frac{1}{|B\cap\ell|}\le\frac{x}{k-1}+\frac{1}{y+1}+(r-x-1).$$

Hence, using the definition of x and y, and the above double sum,

$$\begin{aligned} b&\le w\left(r-\frac{w-1}{k-1}+\frac{1}{y+1}+\frac{y}{k-1}-1\right)\\ &=w\left(r-\frac{w-1}{k-1}\right)+w\left(\frac{1}{y+1}+\frac{y}{k-1}-1\right). \end{aligned}$$

Since $0 \le y \le m-2$, the term $1/(y+1)+y/(k-1)-1$ is ≤ 0. So

$$b-w\left(r-\frac{w-1}{k-1}\right)\le 0.$$

This becomes

$$w^2-w(rk-r+1)+(k-1)b\le 0,$$

which, by substituting $r=(v-1)/(k-1)$ and $b=vr/k$, in turn becomes

$$w^2-wv+\frac{v(v-1)}{k}\le 0.$$

Thus w must lie between the roots of the left-hand side, which implies the claim of the theorem. □

In the first case for which the above result applies, $k=4$, the bound becomes $|B|\ge(v-\sqrt{v})/2$. This does not appear to be a 'best' lower bound in many situations. We leave the reader to compare it with the corresponding affine and projective cases. In terms of existence results for this value of k, we present some results on certain families of $S(2,4,v)$'s. The constructions themselves are beyond the scope of this book, but their detailed investigation would make a good honors thesis project for an interested student.

The constructions of the $S(2,4,v)$'s and their blocking sets are due to Hoffman, Lindner and Phelps (1990, 1991). The reader can refer to Lindner (1990) for a simplified version.

Lemma 8.4.2. *An $S(2,4,v)$ exists if and only if $v\equiv 1$ or 4 (mod 12).*

Proof. We shall prove only that $v\equiv 1$ or 4 is necessary here, because theorem 8.4.3 below claims the existence of $S(2,4,v)$'s for (almost) all such v. Now $r=(v-1)/(k-1)$ (lemma 2.3.3) and $bk=vr$ (exercise 8.7.24), so that $3\,|\,(v-1)$ and $12\,|\,v(v-1)$. Set $v=12t+r$, $0\le r\le 11$. Since $3\,|\,(v-1)$, r can only be 1, 4, 7 or 10. Only $r=1$ or 4 satisfy $12\,|\,v(v-1)$. □

Theorem 8.4.3. *There exists an $S(2,k,v)$ with a blocking set of size $(v-1)/2$ for every $v\equiv 1$ (mod 12), except possibly for $v=37$ and 73. There exists an $S(2,4,v)$ with a blocking set of size $v/2$ for every $v\equiv 4$ (mod 12) except possibly for $v=40$.*

8.5 Blocking sets in generalized quadrangles and partial geometries

We remind the reader that there are some generalized quadrangles that are not partial geometries.

Example 8.5.1. Let S be a grid with line sets partitioned into L_1 and L_2 as in theorem 6.1.1(i). We use this partition to define an $|L_1|\times|L_2|$ matrix and consequently a labelling of the points p_{ij}, $1\le i\le|L_1|$, $1\le j\le|L_2|$. Then the set of points $\{p_{ij}|i+j \text{ even}\}$ forms a blocking set, as does its complement.

In the dual grid, all points of one of the corresponding point sets form a blocking set.

Example 8.5.2. Consider figure 6.1.3, the unique generalized quadrangle on 15 points and 15 lines. The set $\{2,4,6,8,10,12,14\}$ is a blocking set.

Example 8.5.3. The partial geometry of figure 7.1.2 has blocking sets with 5 points.

Since, apart from grids and their duals, generalized quadrangles are special cases of partial geometries, we shall first of all thoroughly examine blocking sets in grids and dual grids before going on to the general case.

Lemma 8.5.1. *Any grid with line class partitions L_1 and L_2 has a blocking set with* $\max\{|L_1|,|L_2|\}$ *points, and this is the minimum size possible. Consequently, in a dual grid, there is always a blocking set with size* $\max\{|P_1|,|P_2|\}$ *where P_1 and P_2 are the point partition sets.*

Proof. We need consider only the grid case. Suppose $|L_1|\ge|L_2|$. Each line of L_1 must meet a blocking set B, giving us $|B|\ge|L_1|$. Since $|L_1|\ge|L_2|$, each point contributed to B from a line of L_1 can be spread over the lines of L_2. Hence the above set of precisely $|L_1|$ points is a blocking set. □

Theorem 8.5.2. *Let B be a blocking set in a partial geometry S with parameters α, s and t, and suppose $\alpha\le s$. Then*

$$|B|\ge\frac{b+(t^2+t+1)+\sqrt{[b-(t^2+t+1)]^2-4b(t-1)}}{2(t+2)}.$$

(Note that if $\alpha=s+1$, we have a Steiner system and can apply theorem 8.4.1.)

Proof. We follow the proof of theorem 8.2.1.

Let x_i, $1\le i\le k$, denote the number of lines of S on exactly i points of B, where $k=\max\{|\ell\cap B|\}$. So

$$\sum_{i=1}^{k}x_i=b=(t+1)(st+\alpha)/\alpha. \qquad \text{(A)}$$

Fix a point $p\notin B$. Then since each line on p meets B, we have $|B|\ge t+1$. Set $|B|=t+\lambda$, $\lambda\ge1$. Counting incidences of points of B with lines of S, we obtain

$$\sum_{i=1}^{k} x_i i = (t+\lambda)(t+1). \tag{B}$$

Counting incidences of ordered pairs of points of B on lines of S, we have

$$\sum_{i=1}^{k} x_i i(i-1) \leq (t+\lambda)(t+\lambda-1). \tag{C}$$

(Note that we can only assume an inequality here, unlike the (C) of theorem 8.2.1.)

Combining as (C)$+\lambda$(B)$-\lambda$(A), we obtain:

$$\sum_{i=1}^{k} x_i(i-1)(i+\lambda) \leq \lambda^2(t+2)+\lambda[t^2+3t-1-b]+t(t-1).$$

Since the left-hand side of this inequality is ≥ 0, we now have

$$0 \leq \lambda^2(t+2)+\lambda[t^2+3t-1-b]+t(t-1).$$

The roots of the right-hand side quadratic in λ are

$$\frac{b+1-3t-t^2 \pm \sqrt{(t^2+3t-1-b)^2-4t(t-1)(t+2)}}{2(t+2)}$$

$$= \frac{b+1-3t-t^2 \pm \sqrt{b(b-2t^2-6t+2)+(t^2+t+1)^2}}{2(t+2)}$$

$$= \frac{b+1-3t-t^2 \pm \sqrt{[b-(t^2+t+1)]^2-4b(t-1)}}{2(t+2)}.$$

So either

$$\lambda+t = |B| \leq \frac{b+(t^2+t+1)-\sqrt{[b-(t^2+t+1)]^2-4b(t-1)}}{2(t+2)}$$

or

$$|B| \geq \frac{b+(t^2+t+1)+\sqrt{[b-(t^2+t+1)]^2-4b(t-1)}}{2(t+2)}.$$

We claim that the first of these inequalities cannot hold. Note first of all that if $|B| < (st+\alpha)/\alpha$, then since each point of B is on $t+1$ lines, we would have $b < (t+1)(st+\alpha)/\alpha$, a contradiction. Hence $|B| \geq (st+\alpha)/\alpha$. Therefore, if the first inequality were true, we would have

$$\frac{st+\alpha}{\alpha} = \frac{b}{t+1} \leq \frac{b+(t^2+t+1)-\sqrt{[b-(t^2+t+1)]^2-4b(t-1)}}{2(t+2)}$$

which is equivalent to

$$\sqrt{[b-(t^2+t+1)]^2-4b(t-1)} \leq (t^2+t+1)-b-\frac{2b}{t+1}.$$

Substituting for b, we see that the right-hand side is negative if and only if

$$0<(s-\alpha)t^2+3st+2\alpha,$$

which is true since $\alpha \leq s$. But this gives a contradiction. □

Corollary 8.5.2. *In case some line of S has at least two points in a blocking set B, then the inequalities of the previous theorem can be improved to strict inequalities.*

Proof. If all lines have precisely one point in B, then the inequality of the last proof corresponding to (C)+λ(B)$-\lambda$(A) has only zero in the sum on the left-hand side. Otherwise, it is strictly bigger than zero. (See exercise 8.7.31.) □

Example 8.5.4. Suppose S is the generalized quadrangle with $v=b=15$, $s=t=2$. Then the above result gives the bound $|B|\geq 3$ for a blocking set B, clearly far away from the actual lower bound.

Example 8.5.5. Consider the partial geometry of figure 7.1.1. Here $t=1$, $s=2$, $v=6$, $b=4$ and $\alpha=2$. The bound of theorem 8.5.2 gives $|B|\geq \frac{4}{3}$ and so $|B|\geq 2$. In fact this partial geometry has blocking sets with just two (non-collinear) points.

In the next lemma, we consider the possibility of a blocking set which has no two points collinear.

Lemma 8.5.3. *Let B be a set of* $(st+\alpha)/\alpha$ *pairwise non-collinear points of a partial geometry S. Then every line of S meets B in a unique point.*

Proof. Each point of B is on $t+1$ lines and none of these is on two or more points of B. Hence,

$$b \geq (t+1)(st+\alpha)/\alpha.$$

Since the right-hand side equals b, it follows that each line meets B precisely once. □

We call a set of $(st+\alpha)/\alpha$ pairwise non-collinear points in a partial geometry an *ovoid*.

Technically, we have excluded grids from this definition, but in view of lemma 8.5.3, we could define an ovoid in a grid to be a set of pairwise non-collinear points such that each line meets it in a unique point. However, it is easy to see that in order for a grid to have an ovoid, all lines must have the same size; also, all grids with constant line size do have ovoids.

In fact, grids with constant line size have the very interesting property that the point set can be partitioned into ovoids. So such a grid would have $s+1$ ovoids. We leave this as exercise 8.7.34.

Lemma 8.5.4. *The union of k disjoint ovoids in a partial geometry, $1 \le k \le s+1$, is a blocking set and each line meets the set in precisely k points.*

Proof. This is clear. Note that k must be $\le s+1$. □

The existence of ovoids is for the most part still open, although partial results are available. We refer the reader to Thas and Payne (1994) for more information.

The property of having ovoids does, however, get passed down to subgeometries in some cases, as we show in the next lemma.

Lemma 8.5.5. *Let $S' = (P', L')$ be a subgeometry of the partial geometry $S = (P, L)$ and suppose $s = s'$. Let O be an ovoid in S. Then $O \cap S'$ is an ovoid in S'.*

Proof. Points of $O \cap S'$ remain pairwise non-collinear in S'. Also, each point of $O \cap S'$ is on $t'+1$ lines of S'. Since $s = s'$, every line of S restricted to P' has the same points in both S and S'. Since each line of S meets O, it follows that each line of S' meets O. Therefore $b' = |O \cap S'|(t'+1)$ and so $|O \cap S'| = (s't'+\alpha')/\alpha'$ implying that O is an ovoid in S'. □

We leave the reader to consider what can happen in case $s' < s$ in exercise 8.7.35.

8.6 Applications of blocking sets

Although the mathematical idea of 'blocking' has been around since the early 1990's, it appeared at first in the context of topology and set theory, and so dealt with infinite sets. The concept arose again, independently, and in a finite setting, in the 1950's and 1960's when Richardson (1956) and Hoffman and Richardson (1961) introduced what they called 'blocking coalitions' into the theory of games. Di Paola (1966, 1969) was the first to consider the idea in a design theory setting, and since then the literature has mushroomed.

The basic idea behind the use of blocking sets in game theory is that two parties are playing a game and the question arises as to whether or not there will always be a winner. This is equivalent to asking if blocking sets exist in the system. Of course one of the two parties, say a casino, might have constructed the game and be quite happy to have a personal win, while trying to stop the other party (a slot-machine player for instance) from winning. In this case, the theory has been modified to allow what have been called *l-blocking* or *hitting sets,* where the set meets every line and may actually contain whole lines.

A recent article by Berardi and Eugeni (1991) gives a brief history of the game theory setting and introduces new 'games' in which blocking sets play an important role. The survey article by Batten (1994) has over ninety references.

The idea of a blocking set can be very useful in statistical sampling or surveys. One problem that surveyors face is getting people to respond truthfully to 'sensitive' questions. (How much money do you make? Have you smoked marijuana in the last twelve months?) Since, in fact, a surveyor does not need to know an individual's response to these questions, but is looking for the average response over a large population of respondents, an approach using blocking sets (though that term is not used) has been suggested by Kuk (1990). The basic idea is the following. A questionnaire will include a number of questions, some of which relate to the sensitive issues, some not. The questionnaire itself is usually formulated in the structure of a mathematical design to make sure that various items or arrangements of items appear the same number of times. The 'sensitive' issues would then form a blocking set, as the surveyor wants to ensure that each grouping of items contains some of the sensitive questions. Now a person responding to the questionnaire does not answer the questions directly, but simply counts the total numbers of yes and no answers in each grouping he or she would make, and reports these numbers to the surveyor. Given this information from a large number of respondents, the surveyor is able to get a good estimate of the proportion of the population whose *truthful* answers to the *sensitive* questions were yes.

A third application of blocking sets concerns cryptography. Define a *2-blocking set* in a projective plane π (other spaces can also be used) to be a set of points B which meets every line in at least two points. We allow B to contain whole lines in this situation. A *cryptosystem* (see Stinson (1995)) is a 5-tuple (P, C, K, E, D) where P is a finite set of elements called *plaintext,* C is a finite set of elements called *ciphertext,* K is a finite set of elements called *keys,* E is a set of functions from P into C and D a set of functions from $E(P)$ into P such that for each $k \in K$ there is a unique $e_k \in E$ and a unique $d_k \in D$ and $d_k(e_k(x)) = x$ for all $x \in P$. The elements of E are called *encryption functions*; those of D, *decryption functions.* The reader is referred to Stinson's book for some simple examples.

Batten has shown (1997) that a cryptosystem can be built as follows. Let P be the set of lines of $\pi = P(2, q)$ written as the set of (q^2+q+1)-tuples from any fixed incidence matrix I of π with points as columns and lines as rows. Let C be the set of all (q^2+q+1)-tuples over GF(2) with precisely $q+1$ non-zero entries. A key (in K) is a triple (T, τ, σ), where T is a triangle comprising $3q$ points in π, τ is a permutation of the corresponding $3q$ columns of I, and σ is a permutation of the other $(q-1)^2$ columns of I. Note that T is a 2-blocking set in π. For $x \in P$, $e_{T,\tau,\sigma}(x)$ is the element y of G obtained by applying π to the $3q$ entries in x corresponding to T, and σ to the remaining entries. The element $d_{T,\tau,\sigma}(y)$ is obtained by reversing the procedure. More details are available in the paper.

Finally, a classic example of the use of blocking sets occurs in scheduling problems. For instance, suppose that v individuals are to participate on b committees, and, so as to distribute the work load fairly, each individual is on the same number r of committees, and each committee has k members. This leads to a Steiner system, as defined in section 2.1. Suppose, in addition, that some of these individuals are female and some are male, and each committee is to have at least one member of each sex. This is asking for a blocking set in the system.

8.7 Exercises

1. Explain why the number of blocking sets in any near-linear space is always even.
2. Which complete graphs have blocking sets?
3. The game 'noughts and crosses' played by two players on a 3×3 grid with entries 0 or X gives rise to blocking sets in case neither player wins. Formalize this game, giving a point set and a block set, and then compute all possible blocking sets. (Note that the point and block set *do not* form a Steiner system!)
4. Find all blocking sets in the spaces of figures 1.4.1 and 2.1.1.
5. Construct blocking sets in the projective planes of (possibly infinite) order ≥ 3, and in the affine planes of order > 3. (See example 8.1.3.)
6. Find a blocking set in the generalized quadrangle of figure 6.1.3 which is neither the example given in 8.1.4 nor its complement.
7. Show that every grid has a blocking set.
8. Find a blocking set in the sense of Bruen and Thas in $P(3, q)$. Is it a 2-blocking set in $P(3, q)$ in the sense of Mazzocca and Tallini?
9. Let π be a finite projective plane of order n, and B a blocking set in π. Show that $|B| = n + \lambda$ for some $\lambda \geq 1$. Show also that no line contains more than λ points.
10. Find the first three values of n for which the existence of blocking sets is not guaranteed by theorem 8.2.3.
11. Construct blocking sets in $P(2, 5)$ of all possible orders given by theorem 8.2.3.

12. Show that the set X of theorem 8.2.3 is a blocking set in π. (Hint: Use section 4.6 to see that it suffices to show that any line consisting of points of the form $[x, mx+b, 1]$ for m and b both non-squares must meet X. In particular, choose $x=-bm^{-1}$ and show that if m and b are non-squares, then bm and hence bm^{-1} are squares.)
13. In the case $n=p^h$ of theorem 8.2.3, show that no line of π is entirely contained in B_1.
14. In the same case as in exercise 13, show that the orders for the set B and its complement do indeed cover the range indicated.
15. Describe all blocking sets in $A(2,4)$.
16. Show that for all $n \geq 5$, an affine plane of order n has a blocking set with $2n-1$ points. (Hint: start with four points forming two pairs of parallel lines.)
17. Show that if B is a blocking set in an affine plane A and $|B|=n+\lambda$, then no line of A contains more than λ points of B.
18. Let $x_1, \ldots, x_t$ be integers ≥ 2. Show that $\binom{\sum_{i=1}^{t} x_i}{2} > \sum_{i=1}^{t} \binom{x_i}{2}$.
19. Show that in theorem 8.3.3, each line of A has a point in and a point not in the set B.
20. In theorem 8.3.3, explain why B_0 is not a blocking set if $n=4$.
21. In theorem 8.3.3, show in detail that $B_0 \cup X_0$ is a blocking set.
22. The following construction is due to Berardi and Eugeni (1984). Let B be a Baer subplane of the projective plane π of square order n, $n \geq 9$. Let $p \in B$. Then p is on $n-\sqrt{n}$ tangents to B, which we label $\ell_1, \ell_2, \ldots, \ell_{n-\sqrt{n}}$. Consider $A=\pi \backslash \ell_1$. On each line ℓ_i, $2 \leq i \leq n-\sqrt{n}$, choose a point p_i such that at most $n-\sqrt{n}-2$ of these are collinear. Let $S=B \backslash \{p\} \cup \{p_2, \ldots, p_{n-\sqrt{n}}\}$. Show that S is a blocking set in A. What goes wrong when $n=4$?
23. The following subsets of $\{1,2,3,4,5,6,7\}$, which we shall call *blocks*, do not form a linear space. They satisfy the condition that every pair of points is on precisely two blocks. Find a blocking set for this system: $\{1,2,4,7\}, \{1,2,3,5\}, \{2,3,4,6\}, \{3,4,5,7\}, \{1,4,5,6\}, \{2,5,6,7\}, \{1,3,6,7\}$.
24. Confirm that the second example of figure 8.4.1 has no blocking set. (This might best be done by computer!)
25. In any $S(2,k,v)$ show that $bk=vr$.

*26. Construct (or find!) $S(2,4,v)$'s for $v=37, 40, 73$.

27. Characterize the generalized quadrangles in theorem 6.1.1(iii) for which $s=1$ or $t=1$.

28. Find all the blocking sets in the generalized quadrangle of figure 6.1.3.

29. Does the partial geometry of figure 7.1.2 have blocking sets on four points? (See example 8.5.3.)

30. Find all blocking sets in the partial geometries of figure 7.1.3.

31. Examine the situation $\sum_{i=1}^{k} x_i(i-1)(i+\lambda)=0$ in the proof of theorem 8.5.2. When does this happen? Can you give a characterization of such blocking sets?

***32.** In case $\alpha = s+1$, how does the bound in theorem 8.5.2 compare with that in theorem 8.4.1?

33. Does the generalized quadrangle of figure 6.1.3 have ovoids?

34. Show that a grid in which all lines have the same size $s+1$ has $s+1$ pairwise disjoint ovoids.

35. Consider what happens to lemma 8.5.4 if $s' < s$.

***36.** Consider $S(4,5,11)$, which is the unique Steiner system with these parameters. Show that it has no blocking set.

37. Let P be the alphabet $a, b, c, \ldots, x, y, z$. Construct a cryptosystem with P as the plaintext.

38. Take $P(2,2)$ and set up the cryptosystem on it as defined in section 8.6.

Bibliography

Ahrens, R. and Szekeres, G. (1969) On a combinatorial generalization of 27 lines associated with a cubic surface. *J. Austral. Math. Soc.* **10**, 485–92.

André, J. (1955) Uber Perspektivitäten in endlichen projektiven Ebenen. *Arch. Math.* **6**, 29–32.

Artin, E. (1957) *Geometric Algebra*. Interscience, New York.

Artin, E. (1971) *Galois Theory*. University of Notre Dame Press, London.

Baer, R. (1946*a*) Polarities in finite projective planes. *Bull. Am. Math. Soc.* **52**, 77–93.

Baer, R. (1946*b*) Projectivities with fixed points on every line of the plane. *Bull. Am. Math. Soc.* **52**, 273–86.

Baer, R. (1947) Projectivities of finite projective planes. *Am. J. Math.* **69**, 653–84.

Baer, R. (1952) *Linear Algebra and Projective Geometry*. Academic Press, New York.

Barlotti, A. (1956) Sui $\{k; n\}$-archi di un piano lineare finito. *Boll. Un. Mat. Ital.* **11**, 553–6.

Barlotti, A. (1965) some topics in finite geometric structures. Lecture notes, Chapel Hill, N.C.

Barlotti, A. (1967) (k, n) arcs in projective planes. *Proc. Projective Geometry Conference*. University of Illinois Press, Chicago, 1–5.

Barlotti, A. (1976) Combinatorics of finite planes and other finite geometric structures. *Foundations of Geometry: Selected Proceedings of a Conference*. University of Toronto Press, 3–15.

Batten, L. M. (1980) Linear spaces with line range $\{n-1, n, n+1\}$ and at most n^2 points. *J. Austral. Math. Soc.* (A) **30**, 215–28.

Batten, L. M. (1994) Blocking sets in designs. *Congr. Numer.* **99**, 139–54.

Batten, L. M. (1997) Protocol for a private key cryptosystem with signature capability based on blocking sets in t-designs. Preprint.

Batten, L. M. and Totten, J. (1980) On a class of linear spaces with two consecutive line degrees. *Ars Combinatoria* **10**, 107–14.

Bell, E. T. (1967) *Development of Mathematics*. McGraw-Hill, New York.

Benson, C. T. (1966) A partial geometry $(q^3+1, q^2+1, 1)$ and corresponding PBIB design. *Proc. Am. Math. Soc.* **17**, 747–9.

Benson, C. T. (1970) On the structure of generalized quadrangles. *J. Algebra* **15**, 443–54.

Berardi, L. and Eugeni, F. (1984*a*) On the cardinality of blocking sets in PG$(2, q)$. *J. Geom.* **22**, 5–14.

Berardi, L. and Eugeni, F. (1984*b*) Blocking sets in affine planes. *J. Geom.* **22**, 167–77.

Berardi, L. and Eugeni, F. (1991) Blocking sets and game theory. *Mitt. Math. Sem. Giessen* **201**, 1–17.

Berman, G. (1952) Finite projective geometries. *Can. J. Math.* **4**, 302–13.

Bernstein, F. (1908) Zur Theorie der trigonometrischen Reihen. *Leipz. Ber.* **60**, 325–38.

Beth, T., Jungnickel, D. and Lenz, H. (1986) *Design Theory.* Cambridge University Press, Cambridge, New York, Melbourne.

Beutelspacher, A. (1980) Blocking sets and partial spreads in finite projective spaces. *Geom. Ded.* **9**, 425–49.

Beutelspacher, A. and Delandtsheer, A. (1980) A common characterization of finite projective spaces and affine planes. *Eur. J. Comb.* **2**, 213–19.

Beutelspacher, A. and Eugeni, F. (1986) On n-fold blocking sets. *Annals Discr. Math.* **30**, 31–8.

Beutelspacher, A. and Mazzocca, F. (1987) Blocking sets in infinite projective and affine spaces. *J. Geom.* **28**, 111–16.

Birkhoff, G. (1935) Combinatorial relations in projective geometries. *Ann. Math.* **36**, 743–8.

Birkhoff, G. (1967) *Lattice Theory,* 3rd edition. American Mathematical Society Colloq. Publ. vol. 25.

Blokhuis, A. (1994) On the size of a blocking set in PG$(2, p)$. *Combinatorica* **14**, 111–14.

Blokhuis, A. and Brouwer, A. E. (1986) Blocking sets in Desarguesian projective planes. *Bull. Lond. Math. Soc.* **18**, 132–4.

Blumenthal, L. M. (1961) *A Modern View of Geometry.* Freeman and Company, San Francisco.

Bose, R. C. (1963). Strongly regular graphs, partial geometries, and partially balanced designs. *Pacif. J. Math.* **13**, 389–419.

Bose, R. C. (1972) Geometric and pseudo-geometric graphs $(q^2+1, q+1, 1)$. *J. Geom.* **2**, 75–93.

Bose, R. C., Freeman, J. W. and Glynn, D. G. (1980) On the intersection of two Baer subplanes in a finite projective plane. *Utilitas Math.* **17**, 65–77.

Bose, R. C. and Shrikhande, S. S. (1959) On the falsity of Euler's conjecture about the non-existence of two orthogonal Latin squares of order $4t+2$. *Proc. Nat. Acad. Sci. USA* **45**, 734–7.

Bose, R. C. and Shrikhande, S. S. (1960) On the construction of sets of mutually orthogonal Latin squares and the falsity of a conjecture of Euler. *Trans. Am. Math. Soc.* **95**, 191–209.

Bose, R. C., Shrikhande, S. S. and Parker, E. T. (1960) Further results on the construction of mutually orthogonal Latin squares and the falsity of Euler's conjecture. *Can. J. Math.* **12**, 189–203.

Bouten, M. and de Witte, P. (1965) A new proof of an inequality of Szekeres, de Bruijn and Erdös. *Bull. Soc. Math. Belg.* **17**, 475–83.

Brauer, R. (1965) On finite Desarguesian planes, I. *Math. Z.* **90**, 117–23.

Brauer, R. (1966) On finite Desarguesian planes, II. *Math. Z.* **91**, 124–51.

Brouwer, A. E. (1979) The number of mutually orthogonal Latin squares – a table up to order 10,000. Math. Zent. Amsterdam, ZW 123.

Brouwer, A. E. and Schrijver, A. (1978) The blocking number of an affine space. *J. Comb. Theory* **24**, 251–3.

Brown, J. M. Nowlin (1970) Homologies and elations of finite projective planes. Ph.D. thesis, Harvard University, Cambridge, Mass.

Brown, J. M. Nowlin (1972*a*) Elations and homologies in collineation groups of finite projective planes. *J. Geom.* **2**, 145–59.

Brown, J. M. Nowlin (1972*b*) Homologies in collineation groups of finite projective planes, I. *Math. Z.* **124**, 133–40.

Brown, J. M. Nowlin (1972*c*) Homologies in collineation groups of finite projective planes, II. *Math. Z.* **125**, 338–48.

Brown, J. M. Nowlin (1976) Odd order elations in projective planes of certain even orders. *Proceedings of the 7th Southeastern International Conference on Combinatorics, Graph Theory and Computing*. Utilitas Mathematics, Winnipeg, 191–202.

Brown, J. M. Nowlin (1977) Homologies generated by elations. *Geom. Ded.* **6**, 435–54.

Brown, J. M. Nowlin (1983) Elations of order 3 in projective planes of order 12. *Finite Geometries, Proceedings of a Conference in Honour of T. G. Ostrom*. Marcel Dekker, New York, 61–6.

Bruck, R. H. (1951) Finite nets I. Numerical invariants. *Can. J. Math.* **3**, 94–107.

Bruck, R. H. (1963*a*) Finite nets, II. Uniqueness and imbedding. *Pacif. J. Math.* **13**, 421–57.

Bruck, R. H. (1963*b*) Existence problems for classes of finite projective planes. Lecture notes, Canadian Mathematics Congress, Saskatoon.

Bruck, R. H. (1967) Construction problems of finite projective planes. *Proceedings of a Conference at University of North Carolina*. Chapel Hill, N.C., 426–514.

Bruck, R. H. (1973) *Construction Problems in Finite Projective Spaces: Finite Geometric Structures and Their Applications*. CIME, 107–88.

Bruck, R. H. and Bose, R. C. (1966) Linear representations of projective planes in projective spaces. *J. Algebra* **4**, 117–72.

Bruck, R. H. and Ryser, H. J. (1949) The non-existence of certain finite projective planes. *Can. J. Math.* **1**, 88–93.

Bruen, A. A. (1970) Baer subplanes and blocking sets. *Bull. Am. Math. Soc.* **76**, 342–4.

Bruen, A. A. (1971) Blocking sets in finite projective planes. *SIAM J. Appl. Math.* **21**, 380–92.

Bruen, A. A. (1973) The number of lines determined by n^2 points. *J. Comb. Theory* (A) **15**, 225–41.

Bruen, A. A. (1980) Blocking sets and skew subspaces of projective space. *Can. J. Math.* **32**, 628–30.

Bruen, A. A. and Silverman, R. (1987) Arcs and blocking sets II. *Eur. J. Comb.* **8**, 351–6.

Bruen, A. A. and Thas, J. A. (1982) Hyperplane coverings and blocking sets. *Math. Z.* **181**, 407–9.

de Bruijn, N. G. and Erdös, P. (1948) On a combinatorial problem. *Indag. Math.* **10**, 421–3, and *Nederl. Akad. Wetensch. Proc. Sect. Sci.* **51**, 1277–9.

Buekenhout, F. (1966*a*) Etude intrinsèque des ovales. *Rendic. Mat.* **25**, 1–61.

Buekenhout, F. (1966*b*) Plans projectifs à ovoides pascaliens. *Arch. Math.* **17**, 89–93.

Buekenhout, F. (1966*c*) Ovales et ovales projectifs. *Atti Accad. Naz. Lincei Rendic. Mat.* **40**, 46–9.

Buekenhout, F. (1969*a*) Une caractérisation des espaces affins basée sur la notion de droite. *Math. Z.* **111**, 367–71.

Buekenhout, F. (1969*b*) Ensembles quadratiques des espaces projectifs. *Math. Z.* **110**, 306–18.

Buekenhout, F. (1976) Characterizations of semi quadrics: a survey. *Atti del Convegno Lincei*. Roma, 393–421.

Buekenhout, F. (1978) Les quadriques et leurs généralisations. Université Libre de Bruxelles. Unpublished lecture notes.

Buekenhout, F. (1979) A characterization of polar spaces. *Simon Stevin* **53**, 3–7.

Buekenhout, F. and Deherder, R. (1971) Espaces linéaires finis a plans isomorphes. *Bull. Soc. Math. Belg.* **4**, 348–59.

Buekenhout, F. and Doignon, J.-P. (1978) Géométrie projective. Université Libre de Bruxelles. Unpublished lecture notes.

Buekenhout, F. and Doyen, J. (1975) Do it with points and lines. Université Libre de Bruxelles. Unpublished lecture notes.
Buekenhout, F. and Lefèvre, C. (1974) Generalized quadrangles in projective spaces. *Arch. Math.* **25**, 540–52.
Buekenhout, F., Metz, R. and Totten, J. (1978) A classification of linear spaces based on quadrangles, I. *Simon Stevin* **53**, 31–45.
Buekenhout, F. and Shult, E. (1974) On the foundations of polar geometry. *Geom. Ded.* **3**, 155–70.
Buekenhout, F. and Totten, J. (1979) A classification of linear spaces based on quadrangles, III. *Geom. Ded.* **8**, 423–35.
Cameron, P. (1974) Partial quadrangles. *Q. J. Math. Oxford* (3) **25**, 1–13.
Cameron, P. J., Goethals, J. M. and Seidel, J. J. (1970) Strongly regular graphs derived from combinatorial designs. *Can. J. Math.* **22**, 597–614.
Chowla, S., Erdös, P. and Straus, E. G. (1960) On the maximal number of pairwise orthogonal Latin squares of a given order. *Can. J. Math.* **11**, 204–8.
Chowla, S. and Ryser, H. J. (1950) Combinatorial problems. *Can. J. Math.* **2**, 93–9.
Cofman, J. (1965) Homologies of finite projective planes. *Arch. Math.* **16**, 476–9.
Cofman, J. (1966) On a characterization of finite Desarguesian projective planes. *Arch. Math.* **17**, 200–5.
Corbas, V. (1964) Omomorfismi fra piani proiettivi, I. *Rendic. Mat.* **23**, 316–30.
Cossu, A. (1960) Sulle ovali di un piano proiettivo sopra un corpo finito. *Atti Accad. Naz. Lincei Rendic. Mat.* **28**, 342–4.
Cossu, A. (1961) Su alcune proprietà dei $\{k;n\}$-archi di un piano proiettivo sopra un corpo finito. *Rendic. Mat.* **20**, 271–7.
Coxeter, H. S. M. (1956) The collineation groups of the finite affine and projective planes with four lines through each point. *Abh. Hamburg* **20**, 165–77.
Coxeter, H. S. M. (1968) *Non-Euclidean Geometry,* 5th edition. University of Toronto Press.
Coxeter, H. S. M. (1974) *Projective Geometry,* 2nd edition. University of Toronto Press.
Crapo, H. and Rota, G.-C. (1970) *Combinatorial Geometries.* MIT Press, Cambridge, Mass.
Cronheim, A. (1953) A proof of Hessenberg's theorem. *Proc. Am. Math. Soc.* **4**, 219–21.
Curtis, C. W. (1967) *Linear Algebra.* Allyn and Bacon Inc., Boston.
Debroey, I. (1981) Semi-partial geometries satisfying the diagonal axiom. *J. Geom.* **13**, 171–90.
Debroey, I. and Thas, J. (1978) On semi-partial geometries. *J. Comb. Theory* (A) **25**, 242–50.
De Clerck, F. (1978) Een kombinatorische studie van de eindige partiele meetkunden. Ph.D. thesis, Rijksuniversiteit, Gent.
De Clerck, F. (1979*a*) Partial geometries – a combinatorial survey. *Bull. Soc. Math. Belg.* **31**, 135–45.
De Clerck, F. (1979*b*) The pseudo-geometric and geometric $(t,s,s-1)$-graphs. *Simon Stevin* **53**(4), 301–17.
De Clerck, F., Dye, R. H. and Thas, J. A. (1980) An infinite class of partial geometries associated with the hyperbolic quadric in $PG(4n-1,2)$. *Eur. J. Comb.* **1**, 323–6.
De Clerck, F. and Thas, J. A. (1978) Partial geometries in finite projective spaces. *Arch. Math.* **30**, 537–40.
Delandtsheer, A. (1983) A classification of finite 2-fold Bolya–Lobachevski spaces. *Geom. Ded.* **14**, 375–93.
Dembowski, P. (1960) Some characterizations of finite projective spaces. *Arch. Math.* **11**, 465–9.
Dembowski, P. (1961) Kombinatorische Eigenschaften endlicher Inzidenzstrukturen. *Math. Z.* **75**, 256–70.

Dembowski, P. (1964) Eine Kennzeichnung der endlichen affinen Räume. *Arch. Math.* **15**, 146–54.
Dembowski, P. (1965) Gruppentheoretische Kennzeichnungen der endlichen desarguesschen Ebenen. *Abh. Hamburg* **29**, 92–106.
Dembowski, P. (1967*a*) Berichtigung und Ergänzung zu "Eine Kennzeichnung der endlichen affinen Räume". *Arch. Math.* **18**, 111–12.
Dembowski, P. (1967*b*) Collineation groups containing perspectivities. *Can. J. Math.* **19**, 924–37.
Dembowski, P. (1968) *Finite Geometries.* Springer-Verlag, New York.
Dembowski, P. and Wagner, A. (1960) Some characterizations of finite projective spaces. *Arch. Math.* **11**, 465–9.
Denniston, R. H. F. (1969) Some maximal arcs in finite projective planes. *J. Comb. Theory* **6**, 317–19.
Dixmier, S. and Zara, F. (1980) Etude d'un quadrangle généralisé autour de deux de ses points non liés. Unpublished manuscript.
Dow, S. (1982) Extending partial projective planes. *Proceedings of the 13th South-eastern International Conference on Combinatorics, Graph Theory and Computing.* Utilitas Mathematica, Winnipeg.
Doyen, J. (1967) Sur le nombre d'espaces linéaires non isomorphes de *n* points. *Bull. Soc. Math. Belg.* **19**, 421–37.
Doyen, J. and Hubaut, X. (1971) Finite regular locally projective spaces. *Math. Z.* **119**, 83–8.
Drake, D. (1985) Blocking sets in block designs. *J. Comb. Theory* (A) **40**, 459–62.
Dulmage, A. L., Johnson, D. and Mendelsohn, N. S. (1959) Orthogonal Latin squares. *Can. Math. Bull.* **2**, 211–16.
Ebert, G. L. (1978) Blocking sets in projective spaces. *Can. J. Math.* **30**, 856–62.
Edge, W. L. (1954) Geometry in three dimensions over GF(3). *Proc. R. Soc. Lond.* (A) **222**, 262–86.
Edge, W. L. (1955) 31-point geometry. *Math. Gaz.* **39**, 113–21.
Edge, W. L. (1956) Conics and orthogonal projectivities in a finite plane. *Can. J. Math.* **8**, 362–82.
Edge, W. L. (1965) Some implications of the geometry of the 21-point plane. *Math. Z.* **87**, 348–62.
Fano, G. (1892) Sui postulati fondamentali della geometria proiettiva. *Giornale di Matematiche* **30**, 106–32.
Fano, G. (1937) Osservazioni su alcune 'geometrie finite', I, II. *Atti Accad. Naz. Lincei* **26**, 25–60, 129–34.
Feit, W. and Higman, G. (1964) The non-existence of certain generalized polygons. *J. Algebra* **1**, 114–31.
Fisher, R. A. (1934) The 6×6 Latin squares. *Proc. Camb. Phil. Soc.* **30**, 492–507.
Foulser, D. A. (1962) On finite affine planes and their collineation groups. Thesis, University of Michigan, Ann Arbor.
Freudenthal, H. (1975) Une étude de quelques quadrangles généralisés. *Ann. Mat. Pur. Appl.* **102**(4), 109–33.
Garner, L. E. (1981) *An Outline of Projective Geometry.* North Holland, New York.
Gemignani, M. C. (1971) *Axiomatic Geometry.* Addison-Wesley, Don Mills, Ontario.
Gleason, A. M. (1956) Finite Fano planes. *Am. J. Math.* **78**, 797–807.
Graves, L. M. (1962) A finite Bolyai–Lobachevsky plane. *Am. Math. Monthly* **69**, 130–2.
Gruenberg, K. W. and Weir, A. J. (1967) *Linear Geometry.* Van Nostrand, Princeton/Toronto/London.
Haemers, W. (1981) A new partial geometry constructed from the Hoffman–Singleton graph. *Finite Geometries and Designs: Proceedings of the Second Isle of Thorns*

Conference. London Mathematical Society Lecture Note series 49, Cambridge University Press, 119–27.

Hall, J. I. (1982) Classifying copolar spaces and graphs. *Q. J. Math. Oxford* (2) **22**, 421–49.

Hall, M. (1943) Projective planes. *Trans. Am. Math. Soc.* **54**, 229–77; and correction, **65** (1949), 473–4.

Hall, M. (1945) An existence theorem for Latin squares. *Bull. Am. Math. Soc.* **51**, 387–8.

Hall, M. (1953) Uniqueness of the projective plane with 57 points. *Proc. Am. Math. Soc.* **4**, 912–16; and correction, **5** (1954), 994–7.

Hall, M. (1954) *Projective Planes and Related Topics*. California Institute of Technology, Pasadena.

Hall, M. (1955) Finite projective planes. *Am. Math. Monthly* **62** (7), part II, 18–24.

Hall, M. (1959) *The Theory of Groups*. Macmillan, New York.

Hall, M. (1960) Automorphisms of Steiner triple systems. *IBM J. Res. Develop.* **4**, 460–72.

Hall, M. (1967) *Combinatorial Theory*. Blaisdell, Waltham, Mass.; 2nd edition (1986), Wiley, New York.

Hall, M. (1971) Affine generalized quadrilaterals. *Studies in Pure Mathematics* (ed. L. Mirsky). Academic Press, New York, 113–16.

Hall, M., Swift, J. D. and Killgrove, R. (1959) On projective planes of order nine. *Math. Comp.* **13**, 233–46.

Hall, M., Swift, J. D. and Walker, R. J. (1956) Uniqueness of the projective plane of order eight. *Math. Tables Aids Comput.* **10**, 186–94.

Hanani, H. (1955) On the number of lines and planes determined by *d* points. *Sci. Public. Technion* **6**, 58–63.

Hanani, H. (1960) A note on Steiner triple systems. *Math. Scand.* **8**, 154–6.

Hanani, H. and Schonheim, J. (1964) On Steiner systems. *Israel J. Math.* **2**, 139–42.

Harary, F. (1969) *Graph Theory*. Addison-Wesley, Reading, Mass.

Hardy, G. H. (1925) What is geometry? *Math. Gaz.* **12**, 309–16.

Hartshorne, R. (1967) *Foundations of Projective Geometry*. W. A. Benjamin Inc., New York.

Hering, C. (1979) On the structure of finite collineation groups of projective planes. *Abh. Math. Sem. Univ. Hamburg* **49**, 82–94.

Hering, C. and Kantor, W. M. (1971) On the Lenz–Barlotti classification of projective planes. *Arch. Math.* **22**, 221–4.

Higman, D. (1971) Partial geometries, generalized quadrangles and strongly regular graphs. *Atti del Convegno di Geometria Combinatoria e sua Applicazioni*. Perugia, 265–93.

Hilbert, D. (1962) *The Foundations of Geometry*. Open Court, Illinois.

Hilton, A. J. W. (1976) Dimension in linear spaces. *Proceedings of the Conference at Orsay, France*.

Hilton, A. J. W. (1977) On the Szamkolowicz–Doyen classification of Steiner triple systems. *Proc. Lond. Math. Soc.* **34**, 102–16.

Hirschfeld, J. W. P. (1979) *Projective Geometries over Finite Fields*. Clarendon Press, Oxford.

Hoffman, A. J. (1965) On the line graph of a projective plane. *Proc. Am. Math. Soc.* **16**, 297–302.

Hoffman, A. J., Newman, M., Straus, E. G. and Taussky, O. (1956) On the number of absolute points of a correlation. *Pacif. J. Math.* **6**, 83–96.

Hoffman, A. J. and Ray-Chaudhuri, D. K. (1965) On the line graph of a finite affine plane. *Can. J. Math.* **17**, 697–94.

Hoffman, A. J. and Richardson, M. (1961) Block design games. *Can. J. Math.* **13**, 110–28.

Hoffman, D. G., Lindner, C. C. and Phelps, K. T. (1990) Blocking sets in designs with block size 4. *Eur. J. Comb.* **11**, 451–7.
Hoffman, D. G., Lindner, C. C. and Phelps, K. T. (1991) Blocking sets in designs with block size four II. *Discr. Math.* **89**, 221–9.
Hubaut, X. (1975) Strongly regular graphs. *Discr. Math.* **13**, 357–81.
Hughes, D. R. (1957) A class of non-Desarguesian projective planes. *Can. J. Math.* **9**, 378–88.
Hughes, D. R. (1959) Collineation groups of non-Desarguesian planes I. The Hall-Veblen-Wedderburn systems. *Am. J. Math.* **81**, 921–38.
Hughes, D. R. (1960) On homomorphisms of projective planes. *Symp. Appl. Math. Proc.* **10**, 45–52.
Hughes, D. R. and Piper, F. C. (1973) *Projective Planes.* Springer-Verlag, New York.
Hughes, D. R. and Piper, F. C. (1985) *Design Theory.* Cambridge University Press, Cambridge, New York, Melbourne.
Jamison, R. E. (1977) Covering finite fields with cosets of subspaces. *J. Comb. Theory* (A) **22**, 253–66.
Kallaher, M. J. (1982) *Affine Planes with Transitive Collineation Groups.* North Holland, New York.
Kantor, W. M. (1969) Characterizations of finite projective and affine spaces. *Can. J. Math.* **21**, 64–75.
Kantor, W. M. (1976) Generalized quadrangles having a prime parameter. *Israel J. Math.* **23**, 8–18.
Kantor, W. M. (1980) Generalized quadrangles associated with $G_2(q)$. *J. Comb. Theory* (A) **29**, 212–19.
Keedwell, A. D. (1965) A search for projective planes of a special type with the aid of a digital computer. *Math. Comp.* **19**, 317–22.
Killgrove, R. B. (1964) Completions of quadrangles in projective planes. *Can. J. Math.* **16**, 63–76.
Killgrove, R. B. (1965) Completions of quadrangles in projective planes, II. *Can. J. Math.* **17**, 155–65.
Knuth, D. E. (1965) A class of projective planes. *Trans. Am. Math. Soc.* **115**, 541–9.
Kuk, A. Y. C. (1990) Asking sensitive questions indirectly. *Biometrika* **77**, 436–8.
Kustaanheimo, P. and Qvist, B. (1954) Finite geometries and their application. *Nordisk Mat. Tidskr.* **2**, 137–55.
Lam, C. W. H., Kolesova, G., and Thiel, L. (1991) A computer search for finite projective planes of order 9. *Discr. Math.* **92**, 187–95.
Lam, C. W. H., Thiel, L., and Swiercz, S. (1991) The non-existence of finite projective planes of order 10. *Can. J. Math.* **41**, 1117–23.
Laskar, R. and Dunbar, J. (1978) Partial geometry of dimension three. *J. Comb. Theory* (A) **24**, 187–201.
Lefèvre-Percsy, C. (1980) Polar spaces embedded in a projective space. *Finite Geometries and Designs: Proceedings of the Second Isle of Thorns Conference.* London Mathematical Society Lecture Note series 49, Cambridge University Press, 216–20.
Lenz, H. (1953) Beispiel einer projektiven Ebene, in der einige, aber nicht alle Vierecke kollineare Diagonalpunkte haben. *Arch. Math.* **4**, 327–30.
Lenz, H. (1954) Zur Begründung der analytischen Geometrie. *Bayerische Akad. Wiss.* **2**, 17–72.
Lenz, H. (1965). Vorlesungen über projektive Geometrie. *Akad. Verl Ges.* Geest & Portig, Leipzig.
Libois, P. (1964) Quelques espaces linéaires. *Bull. Soc. Math. Belg.* **16**, 13–32.
Lindner, C. C. (1990) How to construct a block design with block size four admitting a blocking set. *Australasian J. Comb.* **1**, 101–25.

van Lint, J. H. and Schrijver, A. (1980) Construction of strongly regular graphs, two-weight codes and partial geometries by finite fields. *Combinatorica* **1**, 63–73.

Ljamzin, A. I. (1963) Ein Beispiel eines Paars orthogonaler lateinischer Quadrate der Ordnung 10. *Usp. Math. Nauk* **18**, 173–4.

Lombardo-Radice, L. (1959) *Piani Grafici finite non desarguesiani.* Denaro, Palermo.

Lüneburg, H. (1964) Charakterisierungen der endlichen desarguesschen projektiven Ebenen. *Math. Z.* **85**, 419–60.

Lüneburg, H. (1969) *Lectures on Projective Planes.* University of Illinois, Chicago Circle. Unpublished lecture notes.

Lüneburg, H. (1973) Gruppen und endliche projektive Ebenen. *Finite Geometric Structures and their Applications.* CIME, 213–47.

MacInnes, C. R. (1907) Finite planes with less than eight points on a line. *Am. Math. Monthly* **14**, 171–4.

Mann, H. B. (1942) The construction of orthogonal Latin squares. *Ann. Math. Stat.* **13**, 418–23.

Mann, H. B. (1943) On the construction of sets of orthogonal Latin squares. *Ann. Math. Stat.* **14**, 401–14.

Mann, H. B. (1950) On orthogonal Latin squares. *Bull. Am. Math. Soc.* **50**, 249–57.

Martin, G. E. (1967) On arcs in a finite projective plane. *Can. J. Math.* **19**, 376–93.

Mazzocca, F. (1973) Sistemi grafici rigati di seconda specie. Relazione No. 28. Istituto di Mathematicá dell' Università di Napoli.

Mazzocca, F. and Tallini, G. (1985) On the non-existence of blocking sets in PG(n, q) and AG(n, q) for all large enough n. *Simon Stevin* **59**, 43–50.

Menger, K. (1936) New foundations of projective and affine geometry. *Ann. Math.* **37**, 456–81.

Miller, E. W. (1937) On a property of families of sets. *Comptes Rendus Varsovie* **30**, 31–8.

Moufang, R. (1931) Zur Struktur der projektiven Geometrie der Ebene. *Math. Ann.* **105**, 536–601.

Moulton, F. R. (1902) A simple non-Desarguesian plane geometry. *Trans. Am. Math. Soc.* **3**, 192–5.

Mullin, R. C., Singhi, N. M. and Vanstone, S. A. (1977) Embedding the affine complement of three intersecting lines in a finite projective plane. *J. Austral. Math. Soc.* (A) **24**, 458–64.

Mullin, R. C. and Vanstone, S. A. (1976) A generalization of a theorem of Totten. *J. Austral. Math. Soc.* (A) **22**, 494–500.

Neumann, H. (1955) On some finite non-Desarguesian planes. *Arch. Math.* **6**, 36–40.

Norman, C. M. (1965) Note on the fixed point configuration of collineations. *Math. Z.* **89**, 91–3.

O'Gorman, S. P. (1971) On the generation and automorphisms of projective planes. Ph.D. thesis, London University.

Ore, O. (1962) *Theory of Graphs.* American Mathematical Society, Providence, R.I.

Ostrom, T. G. (1955) Ovals, dualities and Desargues's theorem. *Can. J. Math.* **7**, 417–31.

Ostrom, T. G. (1962*a*) A class of non-Desarguesian affine planes. *Trans. Am. Math. Soc.* **104**, 483–7.

Ostrom, T. G. (1962*b*) Ovals and finite Bolyai–Lobachevsky planes. *Am. Math. Monthly* **69**, 899–901.

Ostrom, T. G. (1965) A characterization of the Hughes planes. *Can. J. Math.* **17**, 916–22.

Ostrom, T. G. (1968) Vector spaces and construction of finite projective planes. *Arch. Math.* **19**, 1–25.

Ostrom, T. G. and Sherk, F. A. (1964) Finite projective planes with affine subplanes. *Can. Math. Bull.* **7**, 549–59.

di Paola, J. W. (1966) On a restricted class of block design games. *Can. J. Math.* **18**, 225–36.
di Paola, J. W. (1969) On minimum blocking coalitions in small projective plane games. *SIAM J. Appl. Math.* **17**, 378–92.
di Paola, J. W. (1985) The shape of minimum blocking sets in small planes. *Ars Comb.* **20**, 15–26.
Parker, E. T. (1959*a*) Construction of some sets of mutually orthogonal Latin squares. *Proc. Am. Math. Soc.* **10**, 946–9.
Parker, E. T. (1959*b*) Orthogonal Latin squares. *Proc. Nat. Acad. Sci. USA* **45**, 859–62.
Parker, E. T. (1962) On orthogonal Latin squares. *Proc. Symp. Pure Math.* **6**, 43–6.
Parker, E. T. (1963) Computer investigations of orthogonal Latin squares of order 10. *Proc. Symp. Appl. Math.* **15**, 73–81.
Parker, E. T. and Killgrove, R. B. (1964) A note on projective planes of order nine. *Math. Comp.* **18**, 506–8.
Payne, S. E. (1970*a*) Affine representation of generalized quadrangles. *J. Algebra* **16**, 473–85.
Payne, S. E. (1970*b*) Collineations of affinely represented generalized quadrangles. *J. Algebra* **16**, 496–508.
Payne, S. E. (1971*a*) Nonisomorphic generalized quadrangles. *J. Algebra* **18**, 201–12.
Payne, S. E. (1971*b*) The equivalence of certain generalized quadrangles. *J. Comb. Theory* (A) **10**, 284–9.
Payne, S. E. (1972) Quadrangles of order $(s-1, s+1)$. *J. Algebra* **22**, 97–119.
Payne, S. E. (1973) Finite generalized quadrangles: a survey. *Proceedings of the International Conference on Projective Planes.* Marcel Dekker, New York, 219–61.
Payne, S. E. (1974) Generalized quadrangles of even order. *J. Algebra* **31**, 367–91.
Payne, S. E. (1975) All generalized quadrangles of order 3 are known. *J. Comb. Theory* (A) **18**, 203–6.
Payne, S. E. (1977*a*) Generalized quadrangles with symmetry, II. *Simon Stevin* **50**, 209–45.
Payne, S. E. (1977*b*) Generalized quadrangles of order 4, I. *J. Comb. Theory* (A) **22**, 267–79.
Payne, S. E. (1977*c*) Generalized quadrangles of order 4, II. *J. Comb. Theory* (A) **22**, 280–8.
Payne, S. E. (1982) Collineations of finite generalized quadrangles. *Finite Geometries, Proceedings of a Conference in Honour of T. G. Ostrom.* Marcel Dekker, New York.
Payne, S. E. (1989) An essay on skew translation generalized quadrangles. *Geom. Ded.* **32**, 93–118.
Payne, S. E. (1990) A census of finite generalized quadrangles. *Finite Geometries, Buildings and Related Topics.* Clarendon Press, Oxford.
Payne, S. E. and Killgrove, R. B. (1978) Generalized quadrangles of order sixteen. *Proceedings of the 9th Southeastern International Conference on Combinatorics, Graph Theory and Computing.* Utilitas Mathematica, Winnipeg, 555–65.
Payne, S. E. and Thas, J. A. (1975) Generalized quadrangles with symmetry. *Simon Steven* **49**, 3–32.
Payne, S. E. and Thas, J. A. (1976*a*) Generalized quadrangles with symmetry, II. *Simon Steven* **49**, 81–103.
Payne, S. E. and Thas, J. A. (1976*b*) Moufang conditions for finite generalized quadrangles. *Finite Geometries and Designs: Proceedings of the Second Isle of Thorns Conference.* London Mathematical Society Lecture Note series 49, Cambridge University Press, 275–303.
Payne, S. E. and Thas, J. A. (1984) *Finite Generalized Quadrangles.* Pitman Press, Boston.

Pickert, G. (1955) *Projektive Ebenen.* Springer-Verlag, Berlin/Göttingen/Heidelberg.
Pickert, G. (1959) Eine Kennzeichnung desarguesscher Ebenen. *Math. Z.* **71**, 99-108.
Pickert, G. (1963) Lectures on projective planes. Notes prepared for Canadian Mathematical Congress, Saskatoon.
Pierce, W. A. (1953) The impossibility of Fano's configuration in a projective plane with eight points per line. *Proc. Am. Math. Soc.* **4**, 908-12.
Piper, F. C. (1963) Elations of finite projective planes. *Math. Z.* **82**, 247-58.
Piper, F. C. (1965*a*) Collineation groups containing elations, I. *Math. Z.* **89**, 181-91.
Piper, F. C. (1965*b*) Collineation groups containing elations, II. *Math. Z.* **92**, 281-7.
Piper, F. C. (1966) On elations of finite projective spaces of odd order. *J. Lond. Math. Soc.* **41**, 641-8.
Piper, F. C. (1967) Collineation groups containing homologies. *J. Algebra* **6**, 256-69.
Piper, F. C. (1968*a*) The orbit structure of collineation groups of finite projective planes. *Math. Z.* **103**, 318-32.
Piper, F. C. (1968*b*) On elations of finite projective spaces of even order. *J. Lond. Math. Soc.* **43**, 459-64.
Primrose, E. J. F. (1951) Quadrics in finite geometries. *Proc. Camb. Phil. Soc.* **47**, 299-304.
Ramamurti, B. (1933) Desargues configurations admitting a collineation group. *J. Lond. Math. Soc.* **8**, 34-9.
Rao, R. C. (1945) Finite geometries and certain derived results in the theory of numbers. *Proc. Nat. Inst. Sci. India* **11**, 136-49.
Ray-Chaudhuri, D. K. (1962) Some results on quadrics in finite projective geometry based on Galois fields. *Can. J. Math.* **14**, 129-38.
Richardson, M. (1956) On finite projective games. *Proc. Am. Math. Soc.* **7**, 458-65.
Rigby, J. F. (1965) Affine subplanes of finite projective planes. *Can. J. Math.* **17**, 977-1014.
Room, T. G. and Kirkpatrick, P. B. (1971) *Miniquaternion Geometry.* Cambridge University Press.
Roth, R. (1964) Collineation groups of finite projective planes. *Math. Z.* **83**, 409-21.
Ryser, H. J. (1955) Geometries and incidence matrices. *Am. Math. Monthly* **62** (7), part II, 25-31.
Ryser, H. J. (1963) *Combinatorial Mathematics.* Wiley, New York.
Sasaki, U. (1952) Lattice theoretic characterization of an affine geometry of arbitrary dimensions. *J. Sci. Hiroshima Univ.* A **16**, 223-38.
Segre, B. (1954) Sulle ovali nei piani lineari finiti. *Atti Accad. Naz. Lincei Rendic.* **17**, 141-2.
Segre, B. (1955) Ovals in a finite projective plane. *Can. J. Math.* **7**, 414-16.
Segre, B. (1957) Sui k-archi nei piani finiti di caratteristica due. *Rev. Math. Pure Appl.* **2**, 289-300.
Segre, B. (1958) Sulle geometrie proiettive finite. *Atti del Convegno Internazionale Reticole e Geometrie Proiettive.* Palermo, 46-61.
Segre, B. (1959*a*) Le geometrie di Galois. *Ann. Mat. Pura Appl.* **48**, 1-96.
Segre, B. (1959*b*) Le geometrie di Galois: Archi ed ovali; calotte ed ovaloidi. *Conf. Sem. Mat. Univ. Bari* **43-4**, 32 pp.
Segre, B. (1959*c*) On complete caps and ovaloids in three-dimensional Galois spaces of characteristic two. *Acta Arithm.* **5**, 315-32.
Segre, B. (1959*d*) Intorno alla geometria di certi spazi aventi un numero finito di punti. *Archimede* **9**, 1-15.
Segre, B. (1960*a*) On Galois geometries. *Proceedings of the International Congress of Mathematicians (Edinburgh).* Cambridge University Press, 488-99.
Segre, B. (1960*b*) Gli spazi grafici. *Rend. Sem. Mat. Fis. Milano* **30**, 223-41.

Segre, B. (1961) *Lectures on Modern Geometry.* Cremonese, Roma.

Segre, B. (1962*a*) Ovali e curve nei-piani di Galois di caratteristica due. *Atti Accad. Naz. Lincei Rendic.* **32**, 785–90.

Segre, B. (1962*b*) Geometry and algebra in Galois spaces. *Abh. Hamburg* **25**, 129–32.

Segre, B. (1964*a*) Arithmetische Eigenschaften von Galois-Räumen, I. *Math. Ann.* **154**, 195–256.

Segre, B. (1964*b*) Teoria di Galois, fibrazioni proiettive e geometrie non desarguesiane. *Ann. Mat. Pura Appl.* **64**, 1–76.

Segre, B. (1965) Istituzioni de geometria superiore. Lecture notes, 3 vols. Univ. di Roma.

Segre, B. (1967) Introduction to Galois geometries. *Lincei Memorie Sc. fisiche* **8**, 137–236.

Seib, M. (1970) Unitäre Polaritäten endlicher projectiver Ebenen. *Arch. Math.* **21**, 103–12.

Seidel, J. J. (1979) Strongly regular graphs. *Surveys in Combinatorics: Proceedings of the Seventh British Combinatorial Conference.* London Mathematical Society Lecture Note series 38, Cambridge University Press, 157–80.

Shrikhande, S. S. (1959) The uniqueness of the L_2 association scheme. *Ann. Math. Stat.* **33**, 781–98.

Shult, E. (1972) Characterizations of certain classes of graphs. *J. Comb. Theory* (B) **13**, 142–67.

Shult, E. (1975) Groups, polar spaces and related structures. *Combinatorics: Proceedings of the NATO Advanced Study Institute.* D. Reidel, Dordrecht, Holland, 451–82.

Skolem, T. (1958) Some remarks on the triple systems of Steiner. *Math. Scand.* **6**, 273–80.

Spencer, J. C. A. (1960) On the Lenz–Barlotti classification of projective planes. *Q. J. Math. Oxford* **11**, 241–57.

Sprague, A. P. (1981) Pasch's axiom and projective spaces. *Discr. Math.* **33**, 79–87.

von Staudt, G. K. C. (1856) *Beiträge zur Geometrie der Lage,* vol. 1. Nürnberg.

Steiner, J. (1853) Combinatorische Aufgabe. *J. Reine Angew. Math.* **45**, 181–2.

Stevenson, F. W. (1970) *Projective Planes.* W. H. Freeman and Company, San Francisco.

Stinson, D. R. (1995) *Cryptography: Theory and Practice.* CRC Press, Boca Raton, London, Tokyo.

Swart, H. and Vedder, K. (1981) Subplane replacement in projective planes. *Aequ. Math.* **22**, 134–9.

Swokowski, E. W. (1979) *Calculus - A First Course,* 2nd edition. Prindle, Weber and Schmidt.

Sylvester, J. J. (1861, 1884) *Collected Papers I and II.* Cambridge University Press.

Szmielew, W. (1983) *From Affine to Euclidean Geometry.* D. Reidel, Boston.

Tallini, G. (1971) Ruled graphic systems. *Atti del Convegno di Geometria Combinatoria e sua Applicazioni.* Perugia, 403–11.

Teirlinck, L. (1975) On linear spaces in which every plane is either projective or affine. *Geom. Ded.* **4**, 39–44.

Thas, J. A. (1971) 4-gonal configurations. *Atti del Convegno di Geometria Combinatoria e sua Applicazioni.* Perugia, 265–93.

Thas, J. A. (1972*a*) 4-gonal subconfigurations of a given 4-gonal configuration. *Lincei-Rend. Sc. fis. mat. nat.* **53**, 520–30.

Thas, J. A. (1972*b*) Ovoidal translation planes. *Arch. Math.* **23**, 110–12.

Thas, J. A. (1973*a*) On 4-gonal configurations. *Geom. Ded.* **2**, 317–26.

Thas, J. A. (1973*b*) Construction of partial geometries. *Simon Stevin* **46**, 95–8.

Thas, J. A. (1974*a*) Construction of maximal arcs and partial geometries. *Geom. Ded.* **3**, 61–4.

Thas, J. A. (1974*b*) On 4-gonal configurations with parameters $r = q^2+1$ and $k = q+1$. *Geom. Ded.* **3**, 365–75.

Thas, J. A. (1975*a*) Some results concerning $\{(q+1)^{(n-1)}; n\}$-arcs and $\{(q+1)(n-1)+1; n\}$-arcs in finite projective planes of order q. *J. Comb. Theory* **19**, 228–32.

Thas, J. A. (1975*b*) 4-gonal configurations with parameters $r = q^2+1$ and $k = q+1$, part II. *Geom. Ded.* **4**, 51–9.

Thas, J. A. (1976) On generalized quadrangles with parameters $s = q^2$ and $t = q^3$. *Geom. Ded.* **5**, 485–96.

Thas, J. A. (1977*a*) Combinatorics of partial geometries and generalized quadrangles. *Higher Combinatorics (Berlin 1976)*. D. Reidel, Dordrecht, Holland, 183–99.

Thas, J. A. (1977*b*) Combinatorial characterizations of the classical generalized quadrangles. *Geom. Ded.* **6**, 339–51.

Thas, J. A. (1978*a*) Partial geometries in finite affine spaces. *Math. Z.* **158**, 1–13.

Thas, J. A. (1978*b*) Combinatorial characterizations of generalized quadrangles with parameters $s = q$ and $t = q^2$. *Geom. Ded.* **7**, 223–32.

Thas, J. A. (1979*a*) Generalized quadrangles satisfying at least one of the Moufang conditions. *Simon Stevin* **53**, 151–62.

Thas, J. A. (1979*b*) Geometries in finite projective and affine spaces. *Surveys in Combinatorics: Proceedings of the Seventh British Combinatorial Conference.* London Mathematical Society Lecture Note series 38, Cambridge University Press, 181–210.

Thas, J. A. (1980*a*) Polar spaces, generalized hexagons and perfect codes. *J. Comb. Theory* (A) **29**, 87–93.

Thas, J. A. (1980*b*) Partial three-spaces in finite projective spaces. *Discr. Math.* **32**, 299–322.

Thas, J. A. and De Clerck, F. (1975) Some applications of the fundamental characterization theorem of R. C. Bose to partial geometries. *Lincei-Rend. Sc. fis. mat. nat.* **59**, 86–90.

Thas, J. A. and De Clerck, F. (1977) Partial geometries satisfying the axiom of Pasch. *Simon Stevin* **51**, 123–37.

Thas, J. A. and Payne, S. E. (1976) Classical finite generalized quadrangles: a combinatorial study. *Ars Combinatoria* **2**, 57–110.

Thas J. A. and Payne, S. E. (1981) Generalized quadrangles and the Higman–Sims technique. *Eur. J. Comb.* **2**, 79–89.

Thas, J. A. and Payne, S. E. (1994) Spreads and ovoids in finite generalized quadrangles. *Geom. Ded.* **52**, 227–53.

Thas, J. A. and De Winne, P. (1977) Generalized quadrangles in projective spaces. *J. Geom.* **10**, 126–37.

Tits, J. (1959) Sur la trialité et certains groups qui s'en déduisent. *Publ. Math. IHES Paris* **2**, 16–60.

Tits, J. (1974) *Buildings of Spherical Type and Finite BN-pairs.* Springer-Verlag, Berlin/Heidelberg/New York.

Totten, J. (1974) Classification of restricted linear spaces. Ph.D. thesis, University of Waterloo.

Totten, J. (1975) Basic properties of restricted linear spaces. *Discr. Math.* **13**, 67–74.

Totten, J. (1976*a*) Classification of restricted linear spaces. *Can. J. Math.* **28**, 321–33.

Totten, J. (1976*b*) Parallelism in a restricted linear space. *Discr. Math.* **14**, 395–8.

Totten, J. (1976*c*) Embedding the complement of two lines in a finite projective plane. *J. Austral. Math. Soc.* (A) **22**, 27–34.

Totten, J. (1976*d*) On the degree of points and lines in a restricted linear space. *Discr. Math.* **14**, 391–4.

Totten, J. (1977*a*) Finite linear spaces with three more lines than points. *Simon Stevin* **51**, 35–47.

Totten, J. (1977*b*) Embeddability of restricted linear spaces. *J. Comb. Theory* **22**, 123–8.
Totten, J. (1979) A classification of linear spaces based on quadrangles, II. *J. Comb. Theory* (A) **27**, 38–49.
Totten, J. and de Witte, P. (1974) On a Paschian condition for linear spaces. *Math. Z.* **137**, 173–83.
Veblen, O. and Bussey, N. J. (1906) Finite projective geometries. *Trans. Math. Soc.* **7**, 241–59.
Veblen, O. and Wedderburn, J. H. M. (1906) Non-Desarguesian and non-Pascalian geometries. *Trans. Math. Soc.* **8**, 379–88.
Veblen, O. and Young, J. W. (1907) *Projective Geometry,* 2 vols. Ginn Co., Boston.
Vedder, K. (1980) Affine subplanes of projective planes. *Finite Geometries and Designs: Proceedings of the Second Isle of Thorns Conference.* London Mathematical Society Lecture Notes series 49, Cambridge University Press, 359–64.
Vedder, K. (1981) A note on the intersection of two Baer subplanes. *Arch. Math.* **37**, 287–8.
Veldkamp, F. D. (1959) Polar geometry, I-V. *Proc. Kon. Ned. Akad. Wet.* **A 62**, 512–51; **A 63**, 207–12. (In *Indag. Math.* **21**, **22**.)
Wagner, A. (1956) On finite non-desarguesian planes generated by 4 points. *Arch. Math.* **7**, 23–7.
Wagner, A. (1959) On perspectivities of finite projective planes. *Math. Z.* **71**, 113–23.
Wagner, A. (1961) On collineation groups of finite projective spaces, I. *Math. Z.* **76**, 411–26.
Walker, A. G. (1947) Finite projective geometry. *Edin. Math. Notes* **36**, 12–17.
Walker, M. (1977) On the structure of finite collineation groups containing symmetries of generalized quadrangles. *Invent. Math.* **40**, 245–65.
Wallis, W. D. (1973) Configurations arising from maximal arcs. *J. Comb. Theory* (A) **15**, 115–19.
Wang, Y. (1964) A note on the maximal number of pairwise orthogonal Latin squares of a given order. *Sci. Sinica* **13**, 841–3.
Whitesides, S. H. (1979) Collineations of projective planes of order 10, parts I and II. *J. Comb. Theory* (A) **26**, 249–68 and 269–77.
Winternitz, A. (1940) Zur Bergründung der projektiven Geometrie: Einführung idealer Elemente unabhängig von der Anordnung. *Ann. Math.* **41**, 365–90.
de Witte, P. (1965) Combinatorial properties of finite planes (in Dutch). Ph.D. thesis, University of Brussels.
de Witte, P. (1966*a*) Combinatorial properties of finite linear spaces, I. *Bull. Soc. Math. Belg.* **15**, 133–41.
de Witte, P. (1966*b*) A new property of non-trivial finite linear spaces. *Bull. Soc. Math. Belg.* **18**, 430–8.
de Witte, P. (1966*c*) Combinatorial properties of finite planes, part I. *Simon Stevin* **40**, 11–19.
de Witte, P. (1967) Some new properties of semi-tactical λ-spaces. *Bull. Soc. Math. Belg.* **19**, 13–24.
de Witte, P. (1973) A new proof of a theorem of Bridges. *Simon Stevin* **47**, 33–8.
de Witte, P. (1975*a*) Restricted linear spaces with a square number of points. *Simon Stevin* **48**, 107–20.
de Witte, P. (1975*b*) Combinatorial properties of finite linear spaces, II. *Bull. Soc. Math. Belg.* **27**, 115–55.
de Witte, P. (1975*c*) The exceptional case in a theorem of Bose and Shrikhande. *J. Austral. Math. Soc.* (A) **24**, 67–78.
de Witte, P. and Batten, L. M. (1983) Finite linear spaces with two consecutive line degrees. *Geom. Ded.* **14**, 225–35.

Wyler, O. (1953) Incidence geometry. *Duke Math. J.* **20**, 601–10.
Yamamoto, K. (1961) Generation principles of Latin squares. *Bull. Inst. Int. Stat.* **38**, 73–6.
Yaqub, J. (1967) The Lenz–Barlotti classification. *Proceedings of the Projective Geometry Conference.* University of Illinois Press, Chicago, 129–62.

Index of notation

Subject index

Printed in the United States
By Bookmasters